Integrating Artificial Intelligence and Machine Learning with Blockchain Security

Integrating Artificial Intelligence and Machine Learning with Blockchain Security

Edited by

D. Jeya Mala and R. Ganesan

Cambridge Scholars Publishing

Integrating Artificial Intelligence and Machine Learning with Blockchain Security

Edited by D. Jeya Mala and R. Ganesan

This book first published 2023

Cambridge Scholars Publishing

Lady Stephenson Library, Newcastle upon Tyne, NE6 2PA, UK

British Library Cataloguing in Publication Data
A catalogue record for this book is available from the British Library

ISBN (10): 1-5275-3011-6
ISBN (13): 978-1-5275-3011-9

We respectfully dedicate this book to all readers who desire to secure software application security through the use of an extensive and knowledgeable risk management system.

TABLE OF CONTENTS

ACKNOWLEDGEMENTS

At the foremost, we would like to thank the Lord Almighty who gave us the strength and courage to complete this book project.

We are deeply grateful to the Chancellor and Management of Vellore Institute of Technology in Chennai for giving the resources necessary to carry out and successfully conclude this project.

We owe a great deal of gratitude to Cambridge Scholars Publishing and the crew for their unwavering support and direction during the writing of the book.

We appreciate every contributor's careful attention to detail and prompt submission of all required documentation. This book will undoubtedly become a reference for researchers all over the world because of their high caliber content.

Last but not least, we would like to thank all the reviewers and proofreaders who contributed to the high caliber of this book.

PREFACE

The notion of Blockchain security has experienced a significant surge in popularity in recent times. The utility of this technology arises from its capacity to construct a reliable and immutable network for conducting financial transactions. The inception of the Blockchain system can be traced back to 2008, when it was originally developed to support the functioning of the digital currency known as Bitcoin. However, it is worth noting that Blockchain technology possesses potential applications that extend beyond the realm of money exchanges. Currently, Blockchain technology is being utilized in a wide range of applications, encompassing areas such as identity verification and supply chain management.

The emergence of Blockchain technology has significantly impacted various aspects of the global landscape. Nevertheless, the vast potential of blockchain technology necessitates a significant level of responsibility, and ensuring the security of blockchain systems is of utmost importance. As the adoption of Blockchain technology increases among enterprises, it becomes imperative to address the concerns pertaining to data security. When making a determination about a Blockchain solution for a corporation, it is imperative to consider a multitude of security measures.

The process of ensuring the security of a Blockchain network include the identification and mitigation of any vulnerabilities that may arise within the distributed ledger of the Blockchain. The design for safeguarding the aforementioned elements is encapsulated in a Blockchain security architecture. The development of a robust security model for Blockchain is crucial in ensuring adequate safeguards for products that rely on this technology.

The administrators of a Blockchain solution require a comprehensive risk model that encompasses all potential threats to the system, encompassing those originating from various organizations, the governance structure, the technology, and the procedures involved. Subsequently, it is important to create a threat model for the Blockchain solution subsequent to a comprehensive evaluation of the associated hazards. In light of the aforementioned insights, there is a pressing need for a book that specifically concentrates on this particular subject matter.

Moreover, the paper titled "Integrating Artificial Intelligence and Machine Learning in Blockchain Security" is specifically focused on the subject of ensuring security inside the realm of Blockchain technology, particularly in relation to its real-world implementations. Each successive chapter of this study examines several application scenarios pertaining to the current advancements in this particular subject. The prioritization of methods that utilize artificial intelligence (AI) and machine learning (ML) for enhancing security in real-time domains is observed.

The chapters have primarily focused on various application domains, including healthcare, supply chain management, computer vision, enterprise resource management, authorization and authentication mechanisms, digital currency, business transactions, operating system level vulnerabilities, internet of things systems security, and fifth-generation wireless communication security. Therefore, this book will serve as a valuable resource for scholars and practitioners seeking to delve deeper into the utilization of artificial intelligence (AI) and machine learning (ML) methodologies in order to bolster the security of Blockchain technology.

CHAPTER 1

BLOCKCHAIN IN HEALTHCARE SECURITY WITH ARTIFICIAL INTELLIGENCE

GAYATHRI R, JOTHY N, GEETHALAKSHMI T
SRI MANAKULA VINAYAGAR ENGINEERING COLLEGE, PUDUCHERRY, INDIA

Abstract

In today's world blockchain is the rapid advancement technology and it provides various opportunities for professionals. As a result, various application sectors are looking for personnel with the competence required to manage this new technology. It is also significant in fields such as healthcare, infrastructure security, digital advertising, and many more. A peer-to-peer network's entire transactions are logged in a blockchain, a type of digital ledger. Participants can validate transactions using this technology without the necessity for a central clearing organization. The major highlights of this chapter are to concentrate in the field of healthcare utilizing blockchain's implications and functions. Blockchain has a major impact on how clinical trial fraud is handled for better patient outcomes. There is a threat in handling the medical documents of patients digitally. With recurring new waves of the pandemic wreaking havoc in the country, more patients prefer to avoid physical touchpoints and are opting to consult doctors online. This section focuses on how Artificial Intelligence can effectively access the blockchain to review the patient's medical data. As a result of integrating the most recent breakthroughs in these technologies, the medical system will have improved service efficiency, cheaper operating costs and decentralized healthcare.

Keywords: Artificial intelligence, blockchain, digital ledger, healthcare, Clinical trial fraud, Patient medical records

Introduction

Blockchain is a decentralised, immutable database that enables asset monitoring and transaction recording across corporate networks. Almost everything of value may be recorded and sold on the blockchain network, minimising risk and increasing efficiency for all parties. Blockchain technology may be integrated into a variety of sectors. Blockchains [1] are mostly utilised as a distributed ledger for monetary transactions. It holds immense promise in a wide range of corporate applications, including banking, finance, government, healthcare, insurance, media and entertainment, retail, and so on.

The advantage of using machine learning (ML) algorithms and other cognitive techniques in the medical field is referred to as "AI" in this context. AI can be easily defined as the ability of computers and other devices to replicate human perception and learn to think, make decisions, and act. A good way to forecast particular results, AI in healthcare frequently refers to the employment of machines to evaluate clinical facts and take action.

Blockchain in e-Health

The rate of development is accelerating in the healthcare sector [2]. There is a desire for superior medical facilities supported by the latest and most advanced technology. This is where blockchain could revolutionise the healthcare industry. Additionally, thc healthcare system landscape is shifting to a patient-centric strategy that emphasises two key elements of these readily available services: adequate medical resources and blockchain technology, which can help healthcare organisations provide adequate patient care and world-class medical facilities. With this technology, you can quickly solve the time-consuming and repetitive process of sharing health information that drives high medical costs. Citizens can participate in health research initiatives using blockchain technology.

The treatment of various groups will also be improved by improved research and shared data on public well-being. The whole healthcare system and groups are managed using a centralised database. Unreliable facts, data protection, and interoperability troubles [3] have historically been the largest challenges in populace fitness care. The blockchain era is dependent on this particular problem. The right use of this generation improves security, record sharing, interoperability, integrity, real-time updates, and access. Statistical safety is also a chief subject, mainly in the areas of the wearable era and personalised medicine. The blockchain era is being used to cope with these problems. This is because sufferers and scientific staff need easy and secure approaches to accumulate, transmit, and query information over networks without demanding approximate safety.

Blockchain security

Blockchain Security is a comprehensive risk management system for blockchain networks, including assurance services, cybersecurity policies, and best practises to mitigate the risk of fraud and cyberattacks.

Blockchain technology data structures have built-in security features such as consensus, encryption, and decentralisation. Each new data block is linked to all the old blocks, making manipulation almost impossible. Additionally, the consensus process (approved users) validates and accepts all transactions, ensuring the truthfulness and accuracy of each transaction. Since users cannot alter transaction records, there is no single point of failure. However, blockchain security goes even further than its built-in security features.

Healthcare will be impacted by blockchain in three key ways:

The word of the year is "blockchain.". As this new technology slowly evolves, it is becoming clear that disruption awaits everything from banking to supply chain operations. The blockchain revolution has a huge opportunity to disrupt and drive digital transformation [4]. There are many ways to use this technology, from medical records to pharmaceutical supply chains and smart contracts to distributing payments.

1. **Health records:** All modern medical systems are based on electronic patient files. But every time you visit the doctor, your medical records lengthen and become more complicated. Healthcare professionals find it difficult to get them because each hospital and doctor's office has a unique method of storing them. Many businesses already exist, including Patient, Medibloc, and the Medical Chain The objective is to offer patients control over their full medical history and to give both patients and doctors one place to access it. In addition to streamlining and improving access, the blockchain would also improve data security [5].

2. **Supply chain:** With one of the highest demands for product safety, protection, and stability, the pharmaceutical business is susceptible to. For example, blockchain technology can track supply chain management securely and openly [6]. This greatly reduces human error and time delays. Additionally, it can be used to track costs, labour, and even waste emissions at any stage of the supply chain. In addition, it can be used to trace products to their origin to confirm their potency and combat the $200 billion annual market loss caused by counterfeit medicines. More effective blockchain logistics solutions are already being developed by businesses like Chronicled, Block Pharma, and Modum. In particular, Modum complies with EU regulations prohibiting pharmaceutical exposure to situations, particularly temperatures, that could impair its quality. The answer provided by Modum was to create a sensor that can capture environmental data while physical goods are being transported and store it permanently on the blockchain.
   - Safeguarding patient data
   - Location of care for genomics administration
   - Management of personal health record data
   - Management of electronic health record data
   - Data management for electronic medical records (EMR)
   - Electronic health records that are interoperable
   - Disease prevention and outbreaks
   - Genomics Protection
      - **Genomic market:** Agencies like EncrypGen and Nebula Genomics are building blockchain platforms that permit people to safely proportion genomic facts in new and rising markets. You're betting that the possibilities surrounding individual genome sequencing will create a billion-dollar data market in the future. Blockchain is the best technology to solve the data security problem and ensure data moves from a source to an end user without These companies target companies that want to use blockchain technology to improve genomic data protection, enable buyers to efficiently collect genomic data, and address the challenges of genomic big data. I'm here. These companies are just a few of the dozens of startups aiming to use blockchain to disrupt healthcare.

Applications of blockchain healthcare

1. Electronic Health Records (EHR):

Over the past ten years, the need for digitising medical records has grown as healthcare professionals and practitioners have become more dependent on quick access to patient data. Electronic health records (EHRs) are digital representations of data that allow authorised healthcare providers to access these records at any time with enhanced security. Electronic health records (EHRs) include patient histories, diagnostic reports, medication information, treatment plans, and test results. Electronic health records are where the blockchain is most frequently used in healthcare systems to increase data security and dependability. Before blockchain technology, the drawback of electronic health records was that patient data was dispersed among different healthcare providers based on the patient's condition, and old data was inaccessible even within EHR systems. Many experts have recommended blockchain as a novel way to store patient EHRs so that their history and present information are safe forever and can be retrieved at any time.

2.Medical supply chain:

One area of the pharmaceutical industry [7] where blockchain might profit from the distinctive qualities and fundamental ideas of technology is the medication supply chain. The pharmaceutical supply chain is considered the most important when implementing technology-driven solutions and use cases. The global market for counterfeit and counterfeit goods is worth up to US$200 billion annually, and counterfeit medicines are a particular problem as the global black market sells such medicines to people without their consent. It is impossible to understate the growing danger that counterfeit drugs and medications pose to human life.

The pharmaceutical business faces serious issues with counterfeiting, which should be looked at as a potential alternative. Blockchain technology has the potential to transform the healthcare supply chain and prevent fraud by improving quality control in the manufacturing and distribution of medicines and drugs. Utilising blockchain technology, it is possible to track the production and distribution of pharmaceuticals as well as the entire lifetime of prescriptions and medications, right down to the individual dose. Several businesses, including Block Pharma, TIERION, CHRONICLED, and the U.S. Centres for Disease Control and Prevention, have adopted blockchain technology for the secure digital labelling of medicines, allowing them to be safely and reliably tracked through their life cycle (CDC).

3. Clinical Research and Development:

Clinical trials involve particularly sensitive aspects of data privacy, integrity, sharing, record keeping, and patient enrollment, and a variety of privacy and security issues could arise. Blockchain, the upcoming internet generation, offers workable solutions to these issues. With the use of blockchain technology, several academics in the field of healthcare are attempting to address the constraints and problems in clinical records.

With the implementation and integration of the blockchain, artificial intelligence (AI), and machine learning techniques, the healthcare sector will undergo even more radical change. Systems to deal with the problem of patient enrolment employ an Ethereum protocol with smart contract capability for clinical data management. They suggested using Ethereum smart contracts for data management system transparency and security in clinical trials and research since they are faster at processing transactions than Bitcoin. Therefore, one of the most widely used current applications of this technology in clinical research and trials is the enrolment of patients via blockchain. A blockchain-based framework was used in another study's implementation to securely track and preserve patients' informed consent. These documents will be safe. These documents will be highly unrealistic, robust, and globally verifiable.

4. Genomic Market:

Genomic data generated by human genomic projects is being heavily utilised in the fields of industry and medical research. However, gene sequencing technologies are also rapidly improving in throughput, giving scientists the ability to speed up the genome sequencing process and reap huge cost savings. By 2025, it is expected that the genomes of 15% of the world's population will be sequenced, generating zettabytes of data. However, no reliable and secure data source can store large amounts of genomic and clinical data. Therefore, there is a need for technologies and tools that can aid in the processing and analysis of genetic data accessible to healthcare professionals, researchers, pharmaceutical companies, and other healthcare providers.

5. Pharmaceutical Medicine

The healthcare industry's fastest-growing and most important area is the pharmaceutical and medical sectors. To give patients the right medication when they need it, the pharmaceutical industry introduces new and potentially effective medications and pharmaceuticals based on current research on a range of patient health concerns [7]. Additionally, it helps to guarantee the legitimacy and safety of medications and other products offered directly to consumers. The pharmaceutical industry also aids in the analysis and preparation of secure medications and pharmaceuticals to hasten patient recuperation. In some typical situations, drug and pharmaceutical corporations are unable to track their medical products promptly because fraudulent or compromised medications may enter the system. The development and distribution of counterfeit drugs and medications via black markets pose a threat to public health on a global scale. Blockchain technology can assess, monitor, and guarantee the security and safety of drug and medical production.

6. Neuroscience Studies

Neuroscience is also an emerging field in healthcare, with new neural techniques exploring new paradigms for controlling data and devices using brain instructions rather than mechanical interactions. The brain's patterns are interpreted by these neural devices, which then convert them into helpful instructions that are utilised to control the external devices. Based on the information they collect from the brain, these gadgets also keep an eye on a person's mental health. Neural interface devices use a variety of sensitive sensors, computation processors, and wireless connectivity to complete the unique task of reading and converting brain impulses. These gadgets interpret electrical signals from your brain and send them to controlled gears. This gear is a device attached to a person's head. All this brain processing is powered by sophisticated algorithms and big data analytics, intended to record these brain signals into a neural interface device using blockchain technology.

In numerous neuroscience applications, including brain augmentation, brain stimulation, and brain thinking, blockchain technology is anticipated to be used as information technology. Blockchain technology offers the capability of brain data storage, which is required for digitising and storing all of the brain's data securely and reliably.

Challenges in Blockchain

The application of blockchain technology [8], which is still in its infancy, is spreading across a variety of industries and has the potential to alter the digital space. However, the technology is still in its infancy and faces some difficulties that will need to be overcome in the deployments to come. Here are some of the significant issues that are covered:

Interoperability: Interoperability is one of the main issues with blockchain technology. Because interoperability allows different users to send and share data and transactions over the network without the need for an intermediary, blockchain design sometimes breaks standards to make the network more flexible. It leads to interoperability and communication issues in blockchain networks.

Uncertainty: The blockchain idea is still in its infancy and cannot be put to use without explicit guidelines and assurance. Currently, a few effective efforts are making use of this contemporary technology. This difficulty must be taken into account if blockchain is to be successfully implemented in ambiguous circumstances.

Storage capacity: The healthcare industry has an enormous amount of medical data, photos, documents, and test results, and storing all these types of data requires a significant amount of storage space. Every node in a blockchain network has a copy of every record, which can exhaust the storage capacity of current blockchain technology.

Cost: Many businesses don't know how much it will cost to set up and maintain healthcare records using blockchain technology, and nobody can use the technology without understanding the precise costs involved.

Use cases for blockchain in the development of digital health

1. Supply chain transparency

Verifying the provenance of pharmaceuticals to validate their efficacy is a major concern not only for the healthcare industry but for many other industries [9]. Using a blockchain-based system to track items at every stage of the supply chain, from the moment they are manufactured, gives customers complete insight and transparency into the products they purchase.

This is a significant problem for the sector, especially in developing countries where tens of thousands of deaths occur each year due to counterfeit prescription drugs. It is also becoming increasingly important for medical devices. Medical devices are rapidly becoming popular with the spread of remote health monitoring and are attracting the attention of malicious attackers.

Blockchain's main advantages, when combined with AI, are:

- **Customer assurance**: The client's ability to follow a package's origins from beginning to end, with integration with producers, wholesalers, transportation.
- **Compliance:** To ensure patient safety, pharmaceutical companies and medical device manufacturers must submit many reports. Combining supply chain data in one system makes it easier to comply with.
- **Supply chain improvement**: Once all data is in one place, companies use AI to better predict demand and optimise supply while preventing criminal intent.

2. Patient-centric electronic health records:

Healthcare systems in every country and region are grappling with the data warehouse problem, which leaves patients and their healthcare professionals with an incomplete picture of their medical history. According to data from Johns Hopkins University released in 2016, medical errors are caused by poorly coordinated care, such as planned actions not being carried out as planned or omissions in a patient's record. It is the third-leading cause of death in the United States.

These are the main advantages of blockchain-enabled EMRs:

- A single source of truth and comprehensiveness for patient medical records improves the experience for both patients and healthcare providers.
- Patients can monitor when their medical records are updated and expressly consent to sharing this data with other individuals or medical professionals. Patients may also limit how long third parties have access to their medical information.
- Allows health insurers to quickly receive verified confirmation of medical benefits from patients without the time and expense of intermediaries.

3. Verifying the credentials of medical staff

Blockchain technology can be used to track the experience of medical professionals and the provenance of medical assets. Reputable healthcare organisations and institutions can log employee credentials, streamlining the recruitment process for these organisations.

The blockchain system's main advantages are:

- Quicker credentialing during the recruiting process for healthcare organisations.
- A chance for medical facilities, insurers, and healthcare providers to monetize their current credential information on former and current employees.
- Assurance and openness to partners.

4. Remote monitoring with IoT security

The use of remote monitoring solutions, in which various sensors track patients' vital signs, is one of the greatest advances in digital health, giving doctors a more accurate picture of a patient's health and enabling more proactive and preventive care. It's one of the trends. A previous article on 5G and edge computing in digital health presented some interesting use cases for remote monitoring.

Artificial Intelligence in e-Health

The vital role that healthcare plays in an affluent and productive society makes healthcare one of the most important industries in a larger big data environment. The use of AI [10–12] in the healthcare industry could mean the difference between life and death. AI can support doctors, nurses, and other healthcare workers in their daily work. AI in healthcare can improve overall patient outcomes, improve prevention and quality of life, and create more accurate diagnostics and treatment strategies.

AI can help predict and monitor infectious disease outbreaks by examining data from the public sector, the healthcare industry, and other sources. As a result, AI could become a key component of global public health efforts to combat pandemics and epidemics. Using ML and other cognitive fields for medical diagnosis is an important application of AI in healthcare.

AI can help doctors and other healthcare professionals use patient data and other information to make more accurate diagnoses and treatment recommendations. Additionally, AI can help make healthcare more proactive and predictable. AI can also help make healthcare delivery more proactive and predictive by analysing big data and suggesting better preventive measures for patients.

Features of AI in e-Health

1. Advanced-Data Analytics

In recent years, the process of gathering data in healthcare [13–15] settings has been streamlined. The data may now be used more successfully in predictive modelling, which will help to improve daily operations and patient care. Instead of concentrating only on historical or current data, we can use both datasets to monitor patterns and generate forecasts.

The practise of supplying healthcare on a price-for-carrier basis is hastily disappearing. In recent years, there has been a tremendous shift towards predictive and preventative interventions in terms of public fitness because of the extended need for patient-centric, or fee-primarily-based, hospital therapy. This is made possible through statistics. Practitioners can identify people who are in high danger of obtaining chronic illnesses and assist in addressing a hassle before it appears instead of just treating the symptoms once they appear. Because preventative care may additionally help to keep off high-priced hospital stays and lengthy-term issues, this lowers costs for the physician, coverage provider, and patient.

2. Chatbots in healthcare

The use of new technology is being driven by emerging trends like expanding service demand, a shift in emphasis toward overall well-being, and rising costs of high-quality care. Medical institutions are changing the rules of patient contact by utilising the power of conversational AI. The healthcare industry is witnessing a rapid increase in the creation and application of numerous AI solutions. The number of technology companies developing AI models and algorithms to improve the quality of healthcare is exploding.

Medical or healthcare chatbots [16] can be used for a variety of purposes, from improving the patient experience and assisting healthcare professionals to streamlining healthcare processes and generating useful insights. Medical

Chatbots are one of the most advanced and impactful AI-powered healthcare solutions ever developed, transforming the way payers, providers, and patients communicate with each other.

AI software tools can detect symptoms, guide medications, and help with long-term health problems. They help patients arrange appointments with specialists and provide appropriate advice for serious illnesses. Over the past decade, chatbot technology has evolved significantly, and both other AI tools have seen tremendous growth.

3. Data Privacy

The application of artificial intelligence (AI) in healthcare is expanding rapidly. The use of AI often revolves around access to medical patient data. As more medical information is shared between patients, doctors, and medical teams via AI products, they are protecting personal information.

The increasing use of AI in healthcare has drawn greater attention to underlying data security and privacy threats, the precautions that have to be taken, and prompted increased oversight and enforcement. Companies that use or sell AI-based healthcare products should comply with the data collection and use rules and regulations governing the retention and use of patient data and other common practicalities. AI-based healthcare products face problems that should be considered.

Radiology Applications

Artificial intelligence in medical diagnosis is not only a powerful tool for minimising physician fatigue but also provides tremendous support for radiology professionals to cope with potentially increasing workloads. To do. Radiologists must work at unprecedented speeds while managing enormous and ever-growing image volumes [17–19]. Today, they must prioritise the most urgent photos, sort through the volume of images, and coordinate patient care.

AI and deep learning solutions can help radiologists manage these high volumes of images by streamlining processes, saving time, increasing capacity, and improving diagnostic accuracy. This considerably lessens the radiologist's workload. Solutions like Aidoc search through enormous amounts of images and flag any that the AI deems to be of concern. The radiologist can then evaluate the highlighted images with the highest priority, identifying urgent cases more quickly without sacrificing their current workload or cases.

Screening for Common Cancers

In routine preventive screenings for malignancies like breast cancer, medical imaging is frequently employed. It can frequently be challenging to definitively determine whether tissue microcalcification in breast cancer is aggressive or benign. False positives result in pointless intrusive testing or treatment, whereas undetected malignancies might cause a delay in diagnosis and worse consequences.

To more effectively classify micro-calcifications by the degree of suspicion for ductal carcinoma in situ (DCIS), AI can help increase accuracy and employ quantitative imaging features, potentially lowering the rate of unnecessary benign biopsies. It can frequently be challenging to definitively determine whether tissue micro-calcification in breast cancer is aggressive or benign. False positives could result in pointless intrusive testing or treatment, whereas undetected malignancies might cause a delay in diagnosis and worse consequences.

Medical Diagnosis

In medical diagnostics, artificial intelligence facilitates medical administration, automation, management, and workflow. Finds acute anomalies, marks significant discoveries in clinical images, supports radiologists in prioritising life-threatening conditions, becomes aware of cardiac arrhythmias, anticipates stroke outcomes, and helps regulate disorders. Rich in data, algorithms, analytics, in-depth knowledge, neural networks, and perceptive machine intelligence, the discipline is constantly growing and changing to meet the demands of the healthcare industry and its customers [20].

AI in drug discovery

AI algorithms can identify uses for new drugs and track their toxicity potential and mechanism of action. This technology has led to the creation of a drug discovery platform that allows the company to repurpose existing pharmaceuticals and bioactive compounds.

The platform's creators may produce more than 80 gigabytes of biological data each week, which is then examined by tools in 1.5 million trials to automate and advance the best facets of biology, data science, and chemical AI research. The potential for human bias is reduced by using machine intelligence methods to extract insights from biological datasets that are too complex for human interpretation. Large pharmaceutical companies are attracted to this strategy because it is less expensive to reuse and redeploy existing drugs than to develop them from scratch [21].

Virtual Care Assistants

Virtual care assistants help the healthcare sector save $20 billion annually by communicating with patients and directing them to the most appropriate care facility. It can monitor patients and respond to queries to provide timely, real-time responses. Most of the virtual care assistant applications available today enable frequent and consistent communication between patients and healthcare professionals. Inadvertent hospitalisations and readmissions are less likely since they happen between patient visits to the clinic. In addition to scheduling doctor visits and tracking patient health, AI-powered virtual assistants can provide patients with a personalized experience to help identify ailments based on symptoms. A virtual nurse will support you during treatment. The AI tool is used to improve patient participation and hone their self-management abilities to prevent serious illnesses from getting worse. The first virtual nurse assistant was Care Angel, which can conduct health checks using voice and AI.

Blockchain and AI Integration in Healthcare

Embedding AI capabilities in the healthcare sector is beneficial, as blockchain can provide security and trust. AI is now mainly used to detect abnormalities in X-rays and CT scans. This is a task that AI can perform at least as effectively as humans and provide more highly personalised treatment [22–26]. Experts predict that important inventions will emerge with huge benefits that can be achieved by reducing the cost of healthcare.

As a result, Google, Microsoft, Apple, Amazon, and many startups are launching new medical applications to make more effective use of patient data, improve certain diagnostic accuracy, etc., and provide relevant recommendations. We are actively investigating AI based on scientific evidence and research results. These apps complement improved surgical care and virtual consultations via mobile apps. Complex algorithms, massive data sets, and efficient machines combine to create a modern approach to integrating medical technology.

Medical records and other data can be transferred to the blockchain via EHR systems. If this data were to be processed directly on the blockchain network, the small block size would increase the amount of computation and memory pressure. Moreover, such data is often subject to invasion of privacy. To solve these challenges, the most important analysis and implementation is to use frameworks to use responsible third parties (e.g., cloud computing) and blockchain for on-chain storage to store large amounts of data. Manage secure original information. Combining next-generation blockchain and artificial intelligence technology to decentralise and accelerate biomedical research and healthcare, deep learning is successfully used in medical diagnosis, cell image analysis, chemical synthesis, and drug classification.

Conclusion

Healthcare workers access blockchain to view patient medical information. AI makes extensive use of data, various proposed algorithms, and decision-making tools. Therefore, healthcare systems will increase service effectiveness, cut costs, and decentralise healthcare by incorporating the most recent innovations into these technologies. Blockchain enables AI to maintain the necessary cryptographic records.

References

1. Abubakar M. et al. 2021 Impact of breast cancer risk factors on clinically relevant prognostic biomarkers for primary breast cancer, Breast Cancer Res Treat:1–13.
2. Adir O, Poley M, Chen G, et al. 2020. Integrating artificial intelligence and nanotechnology for precision cancer medicine, Adv Mater 32(13): 1901989.
3. Al O., Abdullah et al. 2019. A privacy-friendly platform for healthcare data in the cloud based on a blockchain environment, FuturGenerComputSyst 95:511–521.
4. Alotaibi YK, Federico 2017. The impact of health information technology on patient safety, Saudi Med J38 12:1173.

5. Ardila D, Kiraly AP, Bharadwaj S, et al., 2019. End-to-end lung cancer screening with three-dimensional deep learning on low-dose chest computed tomography, Nat Med 25(6): 954–961.
6. Bryatov SR, Borodinov 2019. Blockchain technology in the pharmaceutical supply chain: researching a business model based on hyperledger fabric, CEUR Workshop Proceedings, in, pp. 21–24.
7. Casado-Vara R, Prieto J, la Prieta FD, Corchado 2018. How blockchain improves the supply chain: a case study of the alimentary supply chain, Procedia Comput Sci 134:393–398.
8. Dubovitskaya A, Xu Z, Ryu S, et al. 2017. Secure and trustable electronic medical records sharing using blockchain, AMIA, Annu Symp proceedings, AMIA Symp 2017:650.
9. Durrani S, al-Mushawa F, Heena H, Wani T, and al-Qahtani 2021 Relationship of Oncotype Dx score with tumour grade, size, nodal status, proliferative marker Ki67, and Nottingham Prognostic Index in early breast cancer tumours in the Saudi population, Ann DiagnPathol 51:151674.
10. Khan PW, Byun YC, and Park N (2020b) IoT- PW, Byun YC, and Park N (2020b) IoT-enabled blockchain-optimised provenance systems for the food industry using advanced deep learning, Sensors 20(10):2990.
11. Abadi, Segars, W.P., Tsui, B.M., Kinahan, P.E., Bottenus, N., Frangi, A.F., Maidment, A., Lo, J., and Samei, E., 2020. Virtual clinical trials in medical imaging: a review, Journal of Medical Imaging, 7(4), p. 042805.
12. Adedinsewo D, Carter RE, Attia Z, Johnson P, Kashou AH, Dugan JL, et al. 2020. Artificial Intelligence-Enabled ECG Algorithm to Identify Patients with Left Ventricular Systolic Dysfunction Presenting to the Emergency Department with Dyspnea, Circ Arrhythmia Electrophysiol, 13(8).
13. Adhikari L, Ozrazgat-Baslanti T, Ruppert M, Madushani RWMA, Paliwal S, Hashemighouchani H, et al. 2019. Improved predictive models for acute kidney injury with IDEA: intraoperative data embedded analytics (PLoS Ones)
14. Alsharqi M, Woodward WJ, Mumith JA, Markham DC, Upton R, and Leeson 2018. Artificial intelligence and echocardiography, Echo Res Pract, 5(4):115–25.
15. Aminololama-Shakeri S., Lopez 2019. The Doctor-Patient Relationship with Artificial Intelligence, American Journal of Roentgenology.

16. Andrew D. Selbst, Julia Powles, 2017. Meaningful information and the right to explanation, International Data Privacy Law, Volume 7, Issue 4, Pages 233–242.
17. Caradaica, 2020. Artificial Intelligence and Inequality in the European Union Europolity: Continuity and Change in European Governance [18] Challen, R., Denny, J., Pitt, M., Gompels, L., Edwards, T., and Tsaneva-Atanasova, K., 2019. Artificial intelligence, bias, and clinical safety, BMJ Quality & Safety.
18. De Vries L., Baselmans B., and Bartels M. Smartphone-Based Ecological Momentary Assessment of Well-Being: A Systematic Review and Recommendations for Future Studies, Journal of Happiness 22:2361–2408.
19. Fihn SD, Saria S, Mendonça E, Hain S, Matheny M, Shah N, Liu H, Auerbach A. 2019. "Deploying AI in Clinical Settings." In artificial intelligence in health care: the hope, the hype, the promise.
20. Filice, W., and Ratwani, R.M. 2020. The case for user-centred artificial intelligence in radiology, Radiology: Artificial Intelligence.
21. Forster T., Kentikelenis A., and Bambra, 2018. Health Inequalities in Europe: Setting the Stage for Progressive Policy Action, Foundation for European Progressive Studies.
22. Freeman K, Dinnes J, Chuchu N, Takwoingi Y, Bayliss SE, Matin RN, Jain A, Walter FM, Williams HC, and Deeks JJ. 2020. Algorithm-based smartphone apps to assess the risk of skin cancer in adults: a systematic review of diagnostic accuracy studies.
23. Giulietti M, Cecati M, Sabanovic B, Scirè A, Cimadamore A, Santoni M, et 2021 The role of artificial intelligence in the diagnosis and prognosis of renal cell tumours, Diagnostics, 11(2):206.
24. Golbraikh A., Wang X., Zhu H., and Tropsha A. Predictive QSAR modelling: methods and applications in drug discovery and chemical risk assessment, Handbook of Computational Chemistry.
25. Gómez-González E, Gómez E. 2020 Artificial Intelligence in Medicine and Healthcare: Applications, Availability, and Societal Impact Publications Office of the European Union.
26. Haibe-Kains, 2020. "Transparency and reproducibility in artificial intelligence". Nature, 586(7829):E14–E16.

CHAPTER 2

A SYSTEMATIC REVIEW OF THE INTEGRATION OF BLOCKCHAIN AND KNOWLEDGE GRAPHS IN SUPPLYCHAIN TRANSACTIONS

D. JEYA MALA, SAHIL ARORA

SCHOOL OF COMPUTER SCIENCE AND ENGINEERING,
VELLORE INSTITUTE OF TECHNOLOGY, CHENNAI, INDIA

Abstract

Inefficiencies and disruptions plague the current supply chain. It was fairly obvious during the COVID-19 pandemic. The pandemic exemplified how simple it is to disrupt the entire supply chain. In this scenario, the primary cause of the disorder was a shortage of personnel. Supply chain management refers to the process of optimising the supply chain to reduce transport costs and ensure on-time delivery of the product. Enhancing the supply chain is a crucial aspect of business optimisation. The supply chain can be made more transparent and traceable by utilising blockchain technology. This would be even more crucial in the absence of a central authority based on trust. By means of knowledge graphs, the material transparency throughout the supply chain is increased.The knowledge graphs enable a more comprehensive view of the logistical system. Through this review article, we aim to determine the current state of the literature and identify the system's weaknesses and strengths in order to enhance them.

Keywords: Blockchain, Knowledge Graphs, Supplychain, Transparency, Transactions

Introduction

The contemporary supply chain framework is riddled with interruptions and inadequacies. The outbreak of COVID-19 has illustrated the ease with which the supply chain can be entirely disrupted, resulting in the collapse of entire economies. The implementation of knowledge graphs and blockchain technology is expected to yield the following benefits:

- Enhancing the traceability of goods in the supply chain to ensure adherence to corporate standards
- Mitigating losses arising from counterfeit and grey market trading
- Enhancing visibility and compliance in outsourced contract manufacturing
- Streamlining administrative processes and reducing paperwork

A shipping enterprise engaged in the transportation of commodities across various transit points necessitates the implementation of active tracking, evaluation, and endorsement, resulting in a substantial amount of documentation. A supply chain refers to a network of individuals and organisations that are accountable for the creation and distribution of a commodity to the end-user. The optimisation of the supply chain to reduce transportation costs and ensure timely product delivery is commonly referred to as supply chain management. Enhancing the supply chain is a crucial aspect of optimising business operations. The low cost incurred due to an inefficient supply chain results in delays in the procurement of raw materials and products. The implementation of blockchain technology has the potential to enhance transparency and traceability within the supply chain. This phenomenon presents a prospect for fraudulent activities at various junctures.

The utilisation of knowledge graphs enables the storage of information through the representation of nodes and vertices, which facilitates the navigation of said information through the application of specific algorithms. Additionally, they provide a more extensive viewpoint regarding data that is separated spatially. Our proposed solution involves leveraging blockchain technology and the ledger to expedite this process. This will be achieved through the implementation of smart contracts at every stage, which will automate the management of goods and services. Companies may implement more rigorous management practises when outsourcing contract manufacturing. The implementation of blockchain technology has the potential to mitigate communication or data transfer errors by providing all stakeholders involved in a particular supply chain with equal access to identical information. Efficient delivery of goods and services can be achieved by prioritising quality enhancement over data verification, thereby reducing the time spent on the latter.

This chapter showcases a selection of articles that have been published in recent years. There is a discernible pattern of a rising number of publications pertaining to blockchain and cryptocurrency in academic literature. In addition, there is an increasing amount of scholarly investigation being conducted on topics pertaining to smart contracts, alternative blockchain systems (such as Hyperledger Fabric), and their practical implementations. Despite a decrease in the quantity of articles and publications pertaining to knowledge graphs in recent years,

Search Strategy

The search methodology employed involved utilising specific keywords, namely "Blockchain," "Knowledge Graphs," "Neo4j," "Ethereum," and "Smart Contracts." The articles pertaining to the project's scope were selected and subjected to a meta-review.

Research Question

The primary objective of this chapter is to investigate publications in SCOPUS journals pertaining to the research topic at hand. Through this process, a set of issues were identified for further investigation:

- What measures can be taken to enhance the security of blockchain technology in the context of supply chain management?
- In what ways can the configuration of blockchain technology be optimised to enhance its utility in facilitating B2B transactions?
- What is the extent of publicly available information regarding the supply chain?
- What are the advantages of using graph databases over traditional relational databases?
- What novel concept does it offer that is not offered by relational databases?
- What are the advantages of using smart contracts for data storage in comparison to an on-premises database?
- What are the security implications associated with the implementation of this system?

- What are the maintenance demands and expenses associated with implementing such a system?

Systematic Review

This section presents a systematic review of the existing literature. A comprehensive exploration of pertinent databases was conducted in order to identify noteworthy research studies. The quality of these studies will subsequently be assessed based on predetermined criteria. The present evaluation will provide an up-to-date overview of the existing information on this topic, identify any deficiencies in the literature, and guide forthcoming research.

1. **Information technology of blockchain: database, smart contracts, architecture** (Kuchkovskiy and Shakhovska, 2019) - The subject matter of the article pertains to the emerging field of blockchain technology. The feasibility of developing databases through the implementation of blockchain technology is a topic of discussion. The process of writing data onto the blockchain has been successfully executed in a direct manner. The principal areas of implementation for technology are enumerated. The discourse revolves around the potential of utilising blockchain technology in conjunction with smart contracts. The potential of blockchain technology for future applications is unveiled. The present article discusses the potential of blockchain data analysis in various applications within both research and industry.
2. **Blockchain and supply chain management integration: a systematic review of the literature** (Queiroz and Telles, 2019) - The objective of this research, which scrutinises 106 review papers, is to investigate the current status, potential applications, and forthcoming advancements of blockchain technologies. The exploration of the individuals involved in the development of blockchain technology enhances our comprehension of their endeavours and fortifies their adaptability to forthcoming challenges.
3. **Impact of COVID-19 on logistics systems and disruptions in the food supply chain** (Singh and Kumar, 2021) -The present study discusses the impact of COVID-19 on diverse industries. This study presents a simulation model of the public distribution system (PDS) network that explores three distinct scenarios to demonstrate potential disruptions to the food supply chain. The study

specifically identified the logistical impact and re-routing as areas of concern. This inquiry pertains to the necessary supplementary elements for re-routing protocols and their corresponding management.

4. **Design Procedure of a Knowledge Base for Practical Attack Graph Generation** (Inokuchi and Ohta, 2019) -They developed an attack graph construction system in this study by extending the MulVAL framework, which is renowned for its robust extensibility. In order to generate attack graphs that are of practical utility, a knowledge base was developed and executed within the MulVAL framework, referred to as "interaction rules." As per the assessment, the regulations established by their methodology possess the capability to infer assault scenarios wherein the perpetrator exploits various categories of susceptibilities apart from privilege escalation. Additional rulesets may be incorporated into subsequent research.
5. **Modelling Supply Chain Information Collaboration Empowered with machine learning techniques** (Ali and Ahmed, 2021) -This study uses SVM modelling to study collaborative supply chains. This study effectively addresses the management of multiple business-to-business supply chains, including the handling of various conflicts that may arise due to inconsistent sets of features, such as social media, social events, and location-based store demographic data, in order to produce supply chain results.
6. **Blockchain technology: implications for operations and supply chain management** (Cole and Stevenson, 2019) - The objective of this study is to encourage the investigation of blockchain technology through the lens of operations and supply chain management (OSCM), delineate possible domains of implementation, and outline a roadmap for forthcoming research. This paper serves as a comprehensive appeal to researchers in the field of operations and supply chain management (OSCM) to conduct further inquiries into the potential benefits that blockchain technology can offer to OSCM. The present research fails to address the topic of smart contracts and their potential applications in this context.
7. **Predicting supply chain risks using machine learning: the trade-off between performance and interpretability** (Baryannis and Dani, 2019) - Based on the cooperation of supply chain experts and data-driven AI, this study provides a paradigm for anticipating supply chain hazards. The practicality of the framework is exemplified by means of an empirical investigation of a multi-

layered aerospace production supply chain that is susceptible to the hazard of tardy deliveries. The utilisation of data-driven techniques to extract knowledge and subsequently derive managerial insights is a crucial aspect of influencing the decision-making process in the supply chain.

8. **Smart garbage collection Using GPS and the Shortest Path Algorithm** (Kariapper and Pirapuraj, 2019)-In order to achieve maximum efficiency, this article demonstrates how to utilise GPS and graphs to determine the quickest route for garbage collection through various city neighbourhoods. A graph for calculating the shortest path using GPS was developed and implemented in real-time. The method of path calculation does not account for external factors such as traffic and congestion. The present manuscript exhibits a deficiency in original concepts aimed at enhancing the existing system.
9. **An Alternative Paradigm for Developing and Pricing Storage on Smart Contract Platforms** (Patsonakis and Roussopoulos, 2019) -This paper introduces a new paradigm for building smart contracts, which makes it simpler to validate application data stored on and accessible from an external, perhaps unreliable storage network as their state is always the same size. The discourse pertains to the necessity of standardisation in the development of intelligent contracts and techniques for storing data. The discussion section of this study did not address the matter of scalability.
10. **Design of a Blockchain System for the Protection of Personal Information in a Digital Content Trading Environment** (Heo and Yang, 2020) -This study proposes the SBBC (Secret Block-Based Blockchain) as a potential solution to the challenges associated with current blockchain technology and the digital material sales ecosystem. An algorithm for consensus that is accessible to the public. The aforementioned system has the potential to be implemented on any publicly accessible blockchain. The present research offers a novel perspective on enhancing the existing blockchain technology to facilitate an improved system for trading content.
11. **Power Trading Blockchain using Hyperledger Fabric** (Kim and Kim, 2020) -This study proposes a novel network architecture by incorporating blockchain technology into the existing smart grid system. This study aims to conduct a parametric analysis to compare the transaction throughput and synchronisation overhead of HLF and Ethereum, two prominent blockchain platforms, in

order to evaluate their respective performance. The present research offers a highly specialised illustration wherein the proximity of nodes is physical in nature and the addition of new nodes is infrequent, thereby resulting in challenges related to scalability.

12. **Machine Learning Model for Smart Contract Security Analysis** (Momeni and Wang, 2019) -In this research, the study provided a machine-learning predictive model that finds trends in smart contract security issues. Over 1000 smart contracts that were validated and implemented on the Ethereum platform underwent classification via two customised static code analyzers. Smart contracts can be effectively evaluated by determining their standard issue. The focus of this analysis pertains primarily to code, wherein models are trained to discern the presence or absence of discernible patterns within the solidity code. Instances of false positives are observable.
13. **Community Detection in Scientific Co-Authorship Networks Using Neo4j** (Aung and Nyunt, 2020) -This research investigates the community structure of the coauthor network at the Computer University of Myanmar and employs Neo4j for the purposes of analysis and visualisation. The primary limitation in this scenario pertains to the resolution threshold encountered in instances where the community size is exceedingly diminutive.
14. **The Implementation Knowledge Graph of Air Crash Data Based on Neo4j** (Zou and Liu, 2020) -In addition to discussing how to manage time series data in Neo4j, this study also discusses handling data insertion and creation. The construction of the graph was executed, encompassing the delineation of the interrelationships and the administrative import mechanism. There was no additional implementation subsequent to the creation of the knowledge graph.

Potential Risks of Hyperledger Fabric Smart Contracts (Yamashita and Nomura, 2019) -With the use of static analysis, this study offers a tool for identifying risks. It also describes how to find hazards using our technology and evaluates its value. A tool has been developed that facilitates the identification of 13 distinct types of risks within contractual agreements. The use of general-purpose languages in Hyperledger may give rise to the challenge of composing intrinsically insecure code for contracts.

Results and Discussion

The use of blockchain technology enables any member of a supply chain to exchange verified information without the need for a dependable central authority to serve as an intermediary. According to Kuchkovskiy and Shakhovska (2019, 2-4), the implementation of blockchain technology has the potential to enhance transparency in supply chain management through the instantaneous verification and updating of data. The study was conducted by Queiroz and Telles (2019, 1-3). Blockchain refers to a distributed digital ledger or database that contains records, transactions, or completed events shared among the parties involved. Each transaction within this system is assigned a timestamp and validated through the consensus of the majority of users. Every individual block within a blockchain comprises information pertaining to the transactions executed, the monetary values transacted, and the parties involved. A blockchain is comprised of a sequence of interconnected data blocks that store information.

The blockchain database retains a comprehensive record of all assets and commands executed since the initial one, as the linkage of prior blocks in a chain renders them accessible. Achieving agreement on the implementation of blockchain technology in supply chains that involve multiple layers and partners requires a significant degree of stakeholder engagement. The determination of the risk, viability, and value of blockchain adoption is contingent upon the incentives and the willingness of the involved parties to share information. Blockchain can serve as a valuable tool in situations where a singular source of trust is absent. Despite the availability of alternative measures such as private or permissioned blockchain and strong encryption, the issue of cybersecurity breaches remains a significant concern that must be addressed before the general populace can place their confidence in a blockchain system to safeguard sensitive information.

The implementation of blockchain technology represents a significant advancement in the direction of a decentralised network. Consequently, it is imperative that both its users and operators provide their backing for its continued development and utilisation. Examining the feasibility of utilising blockchain technology for B2B transactions is a significant inquiry. Over time, blockchain technology and smart contracts have demonstrated their durability and applicability in professional contexts, serving as dependable tools for millions of individuals. B2B transactions, which refer to commercial exchanges between two businesses, are often characterised by their larger scale and greater intricacy. There are a greater number of stakeholders that necessitate inclusion within the hierarchical structure of authority.

Stakeholders in the procurement process may comprise procurement teams, end users of the final product, decision-makers, and other relevant parties. The aforementioned circumstances can pose difficulties in pinpointing and directing the appropriate demographic and efficiently conveying the worth of the merchandise or amenities being proffered. The utilisation of blockchain technology holds the capacity to fundamentally transform the business-to-business landscape. The primary advantages of this system are its transparency, inherent security, and increased accountability in transactions. Additionally, it may offer faster processing times compared to traditional systems. The use of an immutable, distributed, and decentralised ledger results in transaction records that are arduous, if not impossible, to modify, thereby mitigating the risks of fraudulent activities and tampering.

Knowledge graphs are used to represent the connections between real-world entities in a structured data format. The ability to amalgamate and organise data from diverse origins in a computationally comprehensible and processable manner is within their purview. The utilisation of knowledge graphs within B2B transactions can facilitate the construction of a comprehensive and cohesive outlook of the various entities implicated, including businesses, products, services, and individuals.

Enhancing the efficiency and precision of transactions and facilitating the discovery of novel associations and perspectives can be achieved through this approach. Chaincode from Hyperledger or smart contracts from Ethereum can be used to establish a blockchain network capable of storing transactions. Determining the superiority of technology, specifically chaincode or smart contracts, is a challenging task as the optimal choice is contingent upon the unique requirements and specifications of the organisation.

The use of smart contracts and chaincode, two technologies designed to automate specific business activities, can enhance the security and dependability of transactions within a blockchain network. Nevertheless, there are notable distinctions between the aforementioned pair. Smart contracts enjoy wider usage and are compatible with multiple blockchain platforms, such as Ethereum. Conversely, chaincode is exclusive to the Hyperledger platform and lacks compatibility with other blockchain technologies.

Additionally, variations in the implementation and execution of chaincode and smart contracts may affect their functionality. One potential use case for smart contracts is the storage of data on the blockchain. The use of smart contracts to record data on the blockchain ensures that the ledger is decentralised and tamper-proof, thereby allowing multiple parties to securely access and verify the information. The utilisation of this technology has the potential to provide benefits in various domains, including but not limited to financial transaction management, legal agreement enforcement, and supply chain information tracking and verification.

Subsequent investigations will focus on various areas, including enhancing the scalability of blockchain systems to facilitate broader acceptance, developing novel algorithms and protocols to enhance the security of blockchain networks, and exploring the probable use cases of blockchain beyond the financial domain, such as in the healthcare and government sectors.

Likewise, certain domains of prospective investigation in knowledge graphs encompass developing more intricate techniques for organising and querying knowledge graphs.

The integration of knowledge graphs with natural language processing (NLP) techniques has been proposed as a means of achieving a more human-like interaction. Additionally, the construction and maintenance of knowledge graphs has been identified as an area of interest, with novel tools and methods being developed to facilitate the integration of data from various sources.

Novel techniques are being devised to improve collaboration between humans and computers by enhancing the visualisation and interaction with knowledge graphs.

The investigation pertains to the utilisation of knowledge graphs for the purpose of advancing explainable artificial intelligence (AI) and guaranteeing the responsibility and transparency of AI systems.

Figure 1 depicts the proportion of research articles included in the review papers that pertain to the subject matter. The ultimate objective was to maintain the central theme of the chapter.

Figure 1 Review papers topic shared in the review

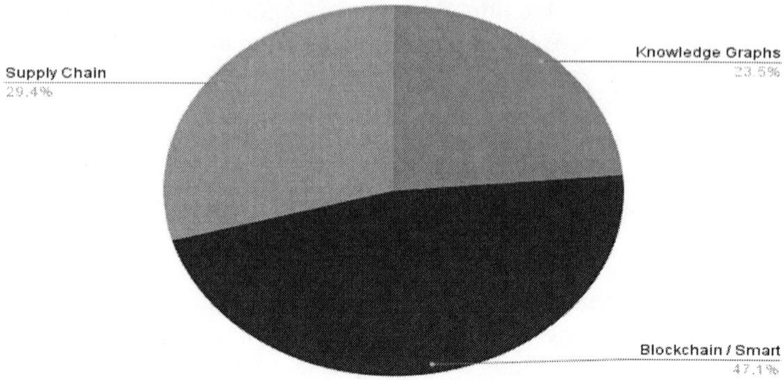

Figure 2 depicts the trend of topics investigated in IEEE journals. The presented graph depicts the temporal evolution of the usage of the terms "blockchain" and "smart contracts" in the titles of research articles published in IEEE papers, spanning the period from 2019 to 2022. There is an observable upward trajectory in the level of attention being paid to these subjects. In the future, it is plausible that an increasing number of routine transactions will be executed through the use of blockchain technology.

The current iteration of blockchain technology is not optimally designed for handling extensive data sets, so scalability represents a significant challenge. Moreover, the use of blockchain as a database may not be universally feasible due to a requisite level of technical proficiency. The management and storage of data constitute a fundamental difference between blockchain technology and traditional databases.

The data in conventional databases is stored in a centralised manner, either on a single server or a network of servers. Consequently, susceptibility to manipulation and censorship arises due to the authority of the party responsible for the database, which enables them to expeditiously modify or eliminate the data. In the context of blockchain, data is distributed across multiple nodes or computers, thereby enhancing its resilience to hacking and censorship.

The level of transparency and reliance represents another noteworthy differentiation. The conventional structure of databases presents a difficulty for users in verifying the accuracy and consistency of the data.

Figure 2 Trend for topics in IEEE articles from 2019 - 2022

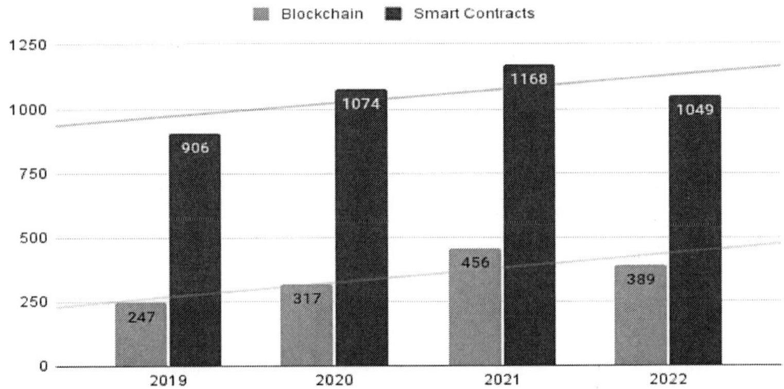

References

1. Ali, N., Ahmed, A., Anum, L., Ghazal, T. M., Abbas, S., Khan, M. A.,...& Ahmad, M. (2021). Modelling supply chain information collaboration empowered with machine learning techniques Tech Science Press, doi: 10.32604/iasc.2021.018983
2. Aung, T. T., &Nyunt, T. T. S. (2020). Community Detection in Scientific Co-Authorship Networks Using Neo4j, 2020 IEEE Conference on Computer Applications (ICCA). doi:10.1109/icca49400.2020.9022826
3. Baryannis, G., Dani, S., & Antoniou, G. (2019). Predicting supply chain risks using machine learning: the trade-off between performance and interpretability Elsevier, Future Generation Computer Systems, 101, 993–1004.
4. Cole, R., Stevenson, M., & Aitken, J. (2019). Blockchain technology: implications for operations and supply chain management Emerald Supply Chain Management: An International Journal. doi:10.1108/scm-09-2018-0309.
5. Heo, G., Yang, D., Doh, I., & Chae, K. (2020). Design of a Blockchain System for Protection of Personal Information in a

Digital Content Trading Environment IEEE 2020 International Conference on Information Networking (ICOIN) doi:10.1109/icoin48656.2020.9016501

6. Inokuchi, M., Ohta, Y., Kinoshita, S., Yagyu, T., Stan, O., Bitton, R.,... Shabtai, A. (2019). Design Procedure of a Knowledge Base for Practical Attack Graph Generation Proceedings of the 2019 ACM Asia Conference on Computer and Communications Security, Asia CCS '19, doi:10.1145/3321705.3329853
7. Kariapper, R., Pirapuraj, P., Razeeth, M. S., Nafrees, A., & Rameez, K. (2019). Smart Garbage Collection Using GPS and the Shortest Path Algorithm 2019 IEEE Pune Section International Conference (PuneCon) doi:10.1109/punecon46936.2019.9105674
8. Kim, Y., Kim, K.-H., & Kim, J.-H. (2020). Power Trading Blockchain using Hyperledger Fabric IEEE 2020 International Conference on Information Networking (ICOIN) doi:10.1109/icoin48656.2020.9016428
9. Kuchkovskiy, &Shakhovska, N. (2019). Information technology of blockchain: database, smart contracts, architecture 2019 IEEE 14th International Conference on Computer Sciences and Information Technologies (CSIT), doi: 10.1109/stc-csit.2019.8929885
10. Momeni, P., Wang, Y., &Samavi, R. (2019, August). Machine learning model for smart contract security analysis. in the 2019 17th International Conference on Privacy, Security, and Trust (PST) (pp. 1–6). IEEE.
11. Patsonakis, C., &Roussopoulos, M. (2019). An Alternative Paradigm for Developing and Pricing Storage on Smart Contract Platforms 2019 IEEE International Conference on Decentralised Applications and Infrastructures (DAPPCON) doi:10.1109/dappcon.2019.00032
12. Queiroz, M.M., Telles, R., & Bonilla, S.H. (2019). Blockchain and supply chain management integration: a systematic review of the literature Emerald, Supply Chain Management, 25, doi: 10.1108/SCM-03-2018-0143
13. Singh, S., Kumar, R., Panchal, R., & Tiwari, M.K. (2021). Impact of COVID-19 on logistics systems and disruptions in the food supply chain. Taylor & Francis Journal, International Journal of Production Research, 59, 1993–2008, doi: 10.1080/00207543.2020.1792000
14. Yamashita, K., Nomura, Y., Zhou, E., Pi, B., & Jun, S. (2019). Potential Risks of Hyperledger Fabric Smart Contracts 2019 IEEE

International Workshop on Blockchain-Oriented Software Engineering (IWBOSE) doi:10.1109/iwbose.2019.8666486

15. Zou, Y., & Liu, Y. (2020). The Implementation Knowledge Graph of Air Crash Data Based on Neo4j*. 2020 IEEE 4th Information Technology, Networking, Electronic, and Automation Control Conference (ITNEC). doi:10.1109/itnec48623.2020.9085182

CHAPTER 3

ANIMAL DETECTION ON ROAD CROSSING IN UNMANNED VEHICLES – A DETAILED STUDY OF APPROACHES IN COMPUTER VISION

[1]SAI SAMYUKTHA N, [2]MALATHI G, [3]BHARADWAJA KUMAR G, [4]GANESAN R

[1]SCHOOL OF COMPUTER SCIENCE AND ENGINEERING, VELLORE INSTITUTE OF TECHNOLOGY, CHENNAI, 600127, TAMIL NADU, INDIA

[2,3,4]PROFESSOR, SCHOOL OF COMPUTER SCIENCE AND ENGINEERING, VELLORE INSTITUTE OF TECHNOLOGY, CHENNAI, 600127, TAMIL NADU, INDIA

Abstract

Road travel is often associated with various risks and uncertainties. For any reason, there could be accidents that result in injuries to the people involved. Among the several causes of road traffic accidents (often referred to as RTAs), the contribution of animals is significant. Hence, while developing an automated, unmanned vehicle, it is important for the vehicle to be well trained with an animal detection model in order to detect moving animals and properly avoid them before a collision (accident) could occur. In this paper, we focus on road accidents due to animals in India. It is a collection of all algorithms and solutions proposed from 2000 to 2022 and includes a detailed study and a proper conclusion with a comparative analysis of all considered solutions and approaches.

Keywords: Animal detection, road traffic accidents, unmanned vehicles, machine learning, computer vision

1. Introduction

There are several causes of road traffic accidents. When roadways are shared by pedestrians and animals with vehicles (irrespective of the speed of movement), accidents increase significantly [1]. The World Road Statistics, 2018 [2], confirms that India ranks first among all 199 countries in the world with regards to the number of road accident deaths. According to the WHO Global Report on Road Safety (2018), India contributes 11% of all deaths due to accidents worldwide [2]. Among all problems related to road safety, animal-vehicle collisions, otherwise known as AVC, are found to be a major threat both for humans and animals [3].

According to the records of the NMC, 3442 and 4262 stray animals were injured during road traffic accidents in the years 2020-21 and 2019-20 respectively [4]. The public usually ignore stray animal accidents due to which statistical figures documented by the NMC official could be comparatively higher in reality. Hence accurate data does not reach the civic body. The causes of most stray animal accidents were found to be not following the speed limit (over speeding) and poor illumination, which led to lesser visibility on the roads. It is recorded that the NGOs working for stray animals receive at least 15 cases daily [4].

According to a study conducted at a tertiary care trauma centre [5] in Eastern India between June 2019 and March 2020, 185 patients experienced road traffic accidents (RTAs) caused by animal-vehicle collisions (AVC). The median age was found to be 29 years, the mode time of accidents (with the highest frequency) was in the evening (4 P.M. to 8 P.M.), and the most frequently occurring vehicles were two-wheelers (92% of cases). The following is found to be the contribution of animals:

- 69% involved stray dogs;
- 21% involved cattle;
- 5% involved oxes; and
- 5% involved other animals.

41% of all RTA victims were diagnosed with Polytrauma, a condition where a person suffers multiple injuries due to an accident which later develops into dysfunction or failure of other organs that hadn't been directly affected during the accident [29]. Patients who were involved in collisions with oxen were found to have more severe injuries (ISS).

The animals found more often on Indian roads are the cow and the dog [6]. Due to the several ways that the animal and vehicle make contact during a collision, the collisions caused by animals can be generally divided into two categories:

a) Direct Collision

In the case of a direct collision, the accident occurs due to direct contact between the animal and the vehicle. Most of the injuries caused here are due to the animal. Commonly, the following three cases of direct collision were noted:

Direct Collision - Case 1

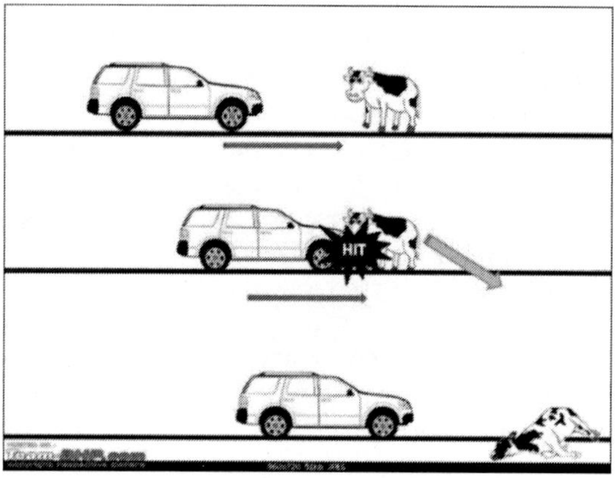

Fig. 1: Direct collision – Case 1

In this case, the collision between the animal and the vehicle results in the animal being thrown away in the forward direction and away from the vehicle. In this case, the vehicle would suffer some damage, but the passengers would be safe (unless there was any secondary impact). This is possible when the collision happens from the side and the force thus produced is sufficient to throw the animal to the side, as is clearly shown in Fig. 1.

Direct Collision - Case 2

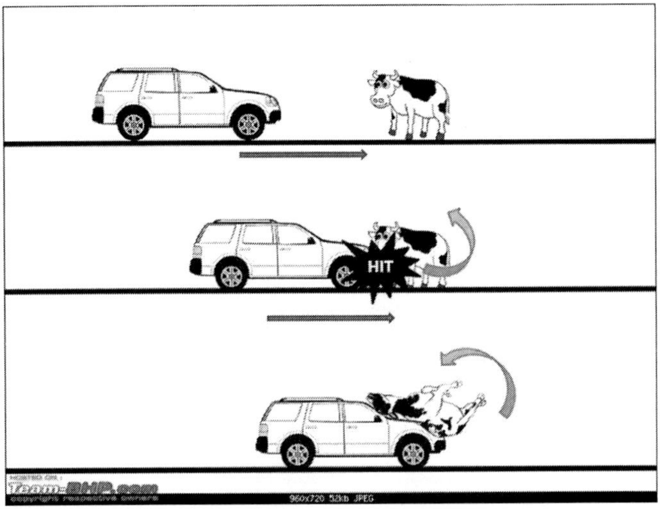

Fig. 2: Direct collision – Case 2

This case may happen when the vehicle is travelling at a considerably higher speed, the impact of which results in the animal being raised in the air and above the vehicle and thrown on the bonnet of the vehicle. This is more serious than Case 1 since there are higher possibilities of the passengers in front and the animal being injured severely, apart from the vehicle. In the worst possible case, there are chances of the front passenger getting stuck between the car and the animal. There is also the possibility that the horns of the animal could injure the passengers in front, as indicated in Fig. 2.

Direct Collision - Case 3

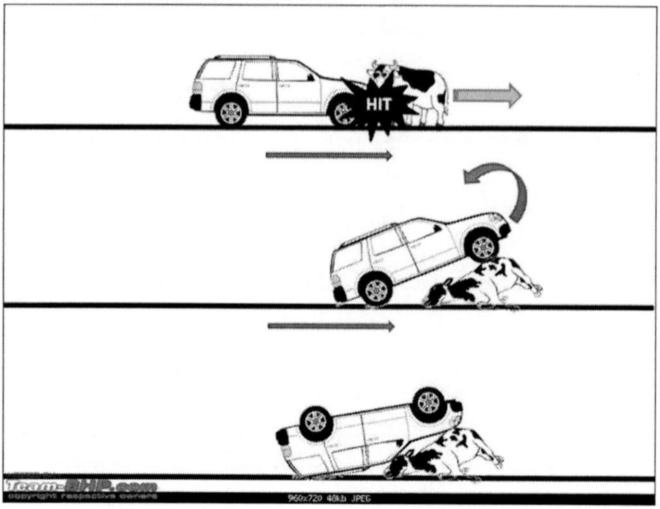

Fig. 3: Direct collision – Case 3

When the vehicle is travelling at a high speed and collision happens, it may result in this case, where the vehicle, after the collision, runs over the animal and gets overturned, as shown in Fig. 3. While at high speed, even after the collision, the car would still have a forward momentum which results in the overturn. This would result in severe injuries to everyone inside the car and also to the animal.

b) Indirect Collision

Indirect Collision - Case 1

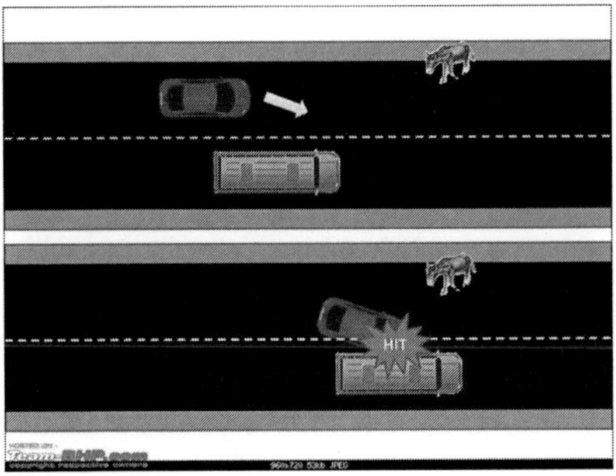

Fig. 4: Indirect Collision - Case 1

In this case, upon noticing the animal crossing, the driver would turn the vehicle to the other side to avoid the collision with the animal. But if there is a moving vehicle in the side lane, it would result in a vehicle-to-vehicle collision and end in an accident, as shown in the Fig. 4.

Indirect Collision - Case 2

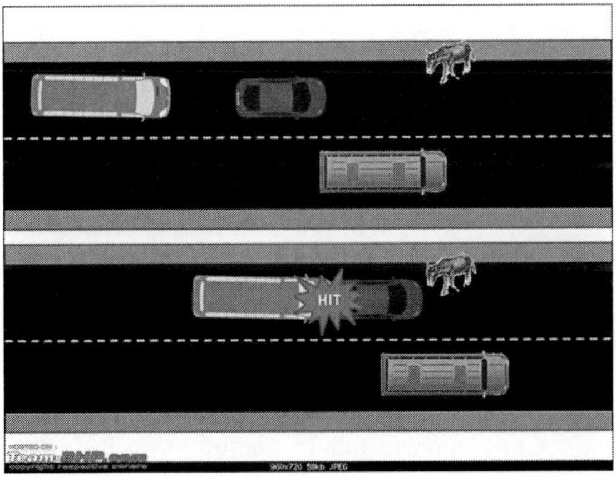

Fig. 5: Indirect Collision - Case 2

In this case, upon noticing the animal crossing, the driver would apply the brakes (often times they are sudden) to slow down and stop the vehicle from colliding with the animal. But if there is a moving vehicle right behind it, this would result in a vehicle-to-vehicle collision as seen in Fig. 5 and end in an accident.

This issue is not specific to India. Records show that across the world, many western countries experience the same issue, with one major difference being that in India, domestic animals cause most of the accidents (cows, oxen, dogs, etc.), whereas wild animals cause accidents in other countries (deer, moose, etc.).

According to statistics, road traffic accidents caused almost 0.11 million deaths in India, accounting for 10% of all road accident fatalities globally. Human factors, vehicles and infrastructure contribute to 92%, 9% and 45% of these deaths respectively. Animals walking on the road cause 6% (12 accidents) of the infrastructure-caused accidents (45%, 91 accidents in total) [7].

Road traffic accidents were found to be more common on state highways as compared to other urban roads [8].

2. Animal Detection Techniques

There are several animal detection techniques proposed to date to prevent animal-vehicle collisions in autonomous vehicles, divided into old approaches (that are ineffective at present) and modern approaches (that are effective at present). Some of the modern approaches are still in the research phase.

A. Old Approaches

I. Direct face detection

According to some researchers in 2006, for an animal to be detected [9], it had to "pose" towards the camera in order to be captured. The approach used here was the one developed by Viola and Jones [10] for the recognition of upright human faces, which originally uses the configurations of the luminance channel's local contrast features. T Burghardt and J Ćalić updated this technique to assess the characteristics of a coloured opponent map by splitting its red and green components.

The resulting map, which was inspired by Troscianko et al. [11], shows the natural landscapes' resistance to the shadows and lighting variations. Here, animal detection is an example of the applications of the proposed algorithm and is not used specifically.

II. Threshold Segmentation approach

The threshold segmentation approach could be used to detect animals [12] in a frame. The goal is to extract information about the target animal from the image's background. However, according to a few other research papers, [13–15], this idea was found to be inefficient since:

- Determining the threshold value is challenging because the background changes frequently.
- This method could also lead to false positives as other dynamic objects could also be detected, such as a moving animal.

III. Power spectrum features approach

A new solution [16] for human action recognition from videos involves the extraction of features in the Fourier domain after fetching the bounding boxes consisting of silhouettes of a human for many such video frames that represent a basic action. Then, certain pre-processing is applied, and each space-time volume is divided into space-time sub-volumes (STSV). Their corresponding mean-power spectra are calculated as feature vectors. This method has high classification performance even with simple distance measures.

Conversely, according to another research paper [17], the experiments conducted by researchers show that detection in 2 out of 3 experiments happened independently of power spectrum differences between image categories. This method is concluded as inefficient due to the computation time required for image processing during its application in real-time.

IV. Animal-specific face detector approach

Another interesting approach to the animal detection domain, as described in the research paper [18], is to use a face detector. Viola and Jones proposed this method which is based on a particular type of animal. Once the animal is detected, it will be tracked over time. But the disadvantage here is that this approach requires the animal to look directly into the camera – any other natural position would not detect the animal.

V. SIFT-based texture descriptor tracking technique

Using a texture descriptor based on SIFT, this approach [19] compares it to a pre-defined library of already-existing animal textures. It uses the "Integral Image," a novel method of image representation that facilitates quick processing. The classifier is developed using the straightforward and efficient AdaBoost learning method.

By doing this, a large number of potential qualities are narrowed down to a select few significant visual aspects. The image's background can be quickly removed using the cascade method, while additional computation is directed toward promising face-like parts. The disadvantage of this technique is that it only works for frames with a single animal and that it is susceptible to background clutter (which should be minimal).

VI. GPS and GPRS based CVAAS system

The rising frequency of camel and car collisions in Saudi Arabia (believed to reach over one hundred annually) prompted the development of the CVAAS employing GPS and GPRS. The Camel-Vehicle Accident-Avoidance System (CVAAS) [20] uses the Global Positioning System (GPS) and General Packet Radio Service (GPRS) to locate camels and then broadcasts the recorded position to the CVAAS server. The CVAAS server evaluates the camel position. If it approaches the danger zone, the camel will signal the drivers by triggering the warning system.

The disadvantage of this strategy is how costly it is to widely install CVAAS. The system produces a lot of false negatives due to its dependence on numerous aspects such as the width of the unsafe zone, variations in camel speed, and delays in SMS message receipt.

VII. Image Mining Algorithm and CBIR Algorithm

The system design [21] consists of web cams positioned in certain detection zones where the animal is more likely to pass by. This is captured as videos on the webcams, which are sent to the processing unit. The processing unit's image mining algorithm highlights any changes to the predetermined reference backdrop. Animal identification using the content-based retrieval technique (CBIR) is applied if any changes are found in the recently collected image. However, the proposed CBIR algorithm has other drawbacks, such as:

- Unsatisfactory querying performance.
- In CBIR systems, a search image's dissimilarity to database images is calculated using distance functions.
- Recovery outcomes are ineffective.
- A very slow approach.
- If the database is enormous, the response time is found to be in the range of minutes.

B. Modern Approaches

I. LIDAR

Researchers created a method using LIDAR (light detection and ranging) to precisely find fishes and distinguish between the water layer, the bottom, and the fishes in the marine environment [22]. For classification purposes, four multi-layer perceptrons (MLP) were created, with the classifications of fish, bottom, and water layer.

II. Animal Detection using Template Matching Algorithm

Template matching [23] identifies small fragments of the input image that match the template. The template matching approach is most frequently used to locate a long-duration signal for a recognised feature, or template.

Template matching is carried out utilising the normalised cross-correlation technique. Cross correlation, in the field of signal processing, refers to the percentage of time slack applied to one of the waveforms that accounts for the degree of similarity between two waveforms, otherwise known as the sliding point product or the sliding inner-product.

In order to determine a picture's brightness, normalisation of images is first performed for applications in the field of image processing techniques; in this case, the template can vary according to the amount of lightning and exposure of the surroundings. Subtracting the mean and dividing by the standard deviation results in Image normalization. This is the feature-based template matching method utilising NCC.

III. Deep Learning - CNN Based

In [23], the focus is on involuntarily and accurately gathering the camera trap image data by collecting the movements of wildlife using motion sensors. Using deep learning, the location and movements of wild animals are already known, as shown in Fig. 6. Information can be extracted from the collected data, but the downside of this is that it is a costly, sustained physical task.

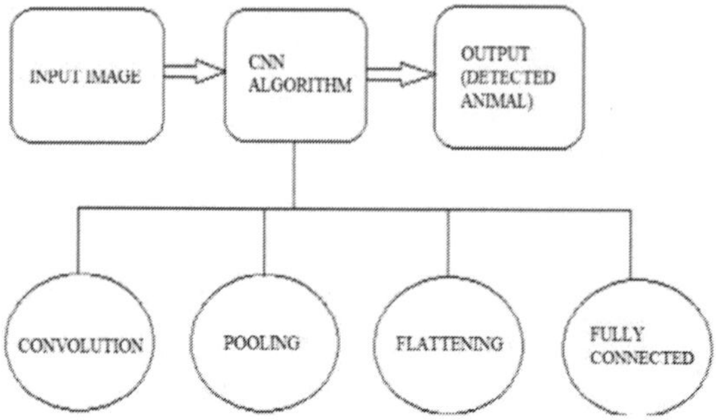

Fig. 6: Block Diagram / Architecture of the Algorithm

A dataset with 3.2 million images from Snapshot Serengeti is used here. The dataset is divided between training and testing groups in a 75:25 ratio.

Convolutional Neural Network

A deep convolutional neural network (CNN) is trained to identify, quantify, and depict 48 breeds' behaviors in the aforementioned dataset. With an accuracy rate of 93.8%, this network can recognise animals automatically. Considering only self-confidant image classification, animal identification is automated for 99.3% of the data.

Natural language processing (NLP) and image processing applications frequently use CNNs (convolutional neural networks). It is a subset of artificial neural networks that analyses vast amounts of data using the perceptron machine learning method for supervised learning.

An input layer, an output layer, and a large number of hidden layers make up the basic framework of CNN. Only a few of these layers pass findings to the following levels using convolved methods [23].

- ***The Input layer:*** obtains the input image that contains the raw pixel values of three primary colours – R (Red), G (Green), B (Blue)
- ***The CONV layer:*** this layer calculates the output connected to local regions to the input layer.

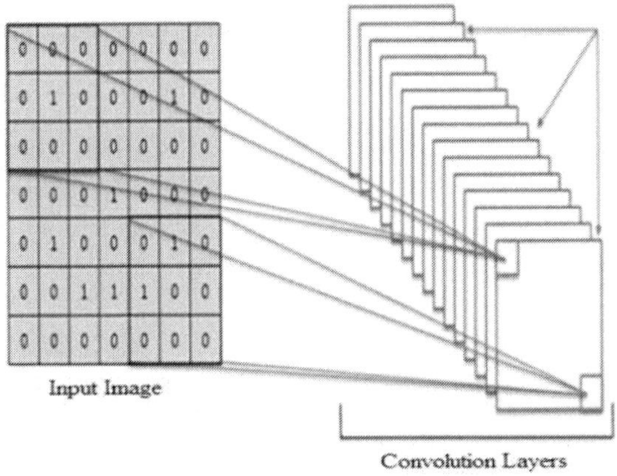

Fig. 7: The Convolutional Layer

In this layer, the features from the input images are extracted as shown in Fig. 7. Here, processing of the input image will happen by the application of a suitable filter to do operations such as edge detection, blur, and image sharpening.

- ***The RELU layer:*** this layer applies a unit-wise activation function, making the size of the volume unaffected.
- ***The POOL layer:*** this layer performs the down sampling procedure and generates 16 x 16 x 2 size along with unique dimensions (width and height).

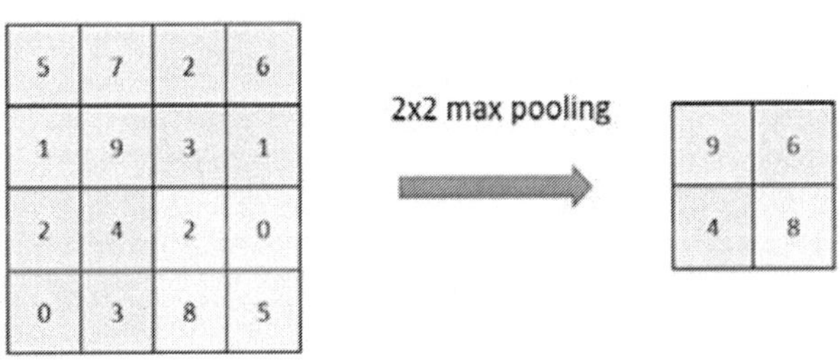

Fig. 8: Pooling layer

Pooling layer decreases the number of parameters in a large input image. Taking the largest element from the revised feature map would result in Max pooling as shown in Fig. 8 – the goal here is to down sample the provided input image by selecting the largest element from the updated feature map.

- ***Flattening***

This procedure is used for converting the output image from convolutional layers in 2D array form into a single, long, and continuous linear vector (1D), as shown in Fig. 9.

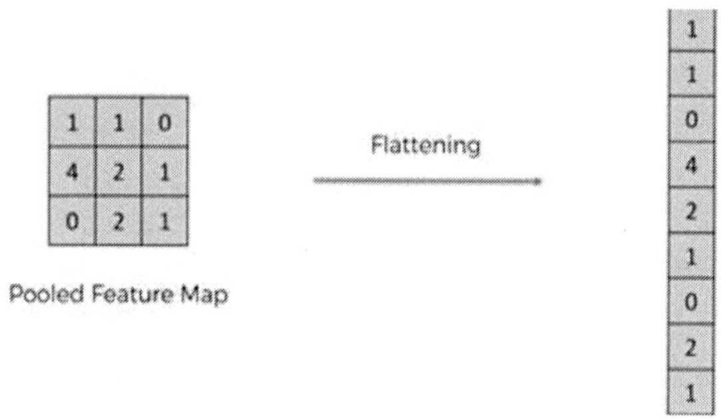

Fig. 9: Flattening layer

- ***FC layer***: this layer will compute the class grades, producing a size-related volume. Similar to traditional neural networks, every neuron in this layer is connected to every weight in the set before it.

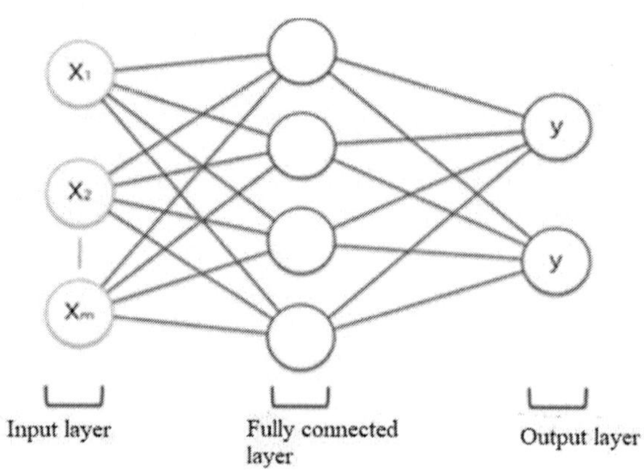

Fig. 10: Fully Connected layer

The Fig. 10 shows all the hidden layers of CNN. For a more accurate prediction of the output, this method integrates traits into multiple features.

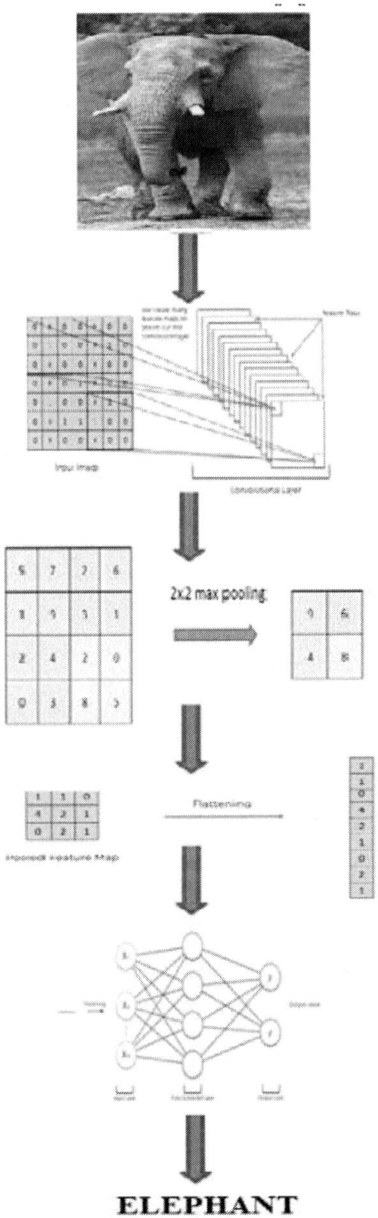

Fig. 11: Flow diagram of the algorithm

Result

This method enables 96.6% accurate, significant, inexpensive, and real-time collections of large numbers of animals in their natural habitats.

Test results when tested against different sets of data are:

- Elephant blurred test image, the accuracy is 0.7925.
- Cheetah blurred test image, the accuracy is 0.8679.

IV. HOG and Cascade Classifier

In Sharma and Shah's paper [7], feature descriptors such as HOG, or Histogram of Oriented Gradients, are used in computer vision applications for the identification of objects in video or picture formats.

Efficiency: Early candidate elimination (false positives) due to the cascading process

The solution proposed focuses on an animal detection algorithm, with the output alerting the driver accordingly.

Architecture of Animal Detection and Collision Avoidance System:

The video captured by the forward-facing camera (optical sensor) attached to the vehicle is first converted into frames of images. The video includes instances of both fixed (stationary) and moving (non-stationary) items, along with animals, on the road. Then, certain pre-processing is applied to clean the image for the next step. Now, the animal detection algorithm is applied, and the distance is calculated based on which the driver would be alerted. The system uses a combination of a histogram of oriented gradients (HOG) and a boosted cascade classifier to extract features and learn from them. The image processing techniques used in this research paper are implemented using OpenCV software. Apart from animal detection, distance calculation is also made here. Based on the obtained values, the driver would take necessary action after displaying the alert message on the command prompt. There are three different messages that could be sent to the driver depending on the distance determined:

Condition	Message
Animal too close to the vehicle	"Animal very near"
Animal is at a considerably far distance from the vehicle	"Animal little far"
Animal too far from the vehicle (at a safe distance)	"Very far"

This process is depicted as a flow diagram in Fig. 12.

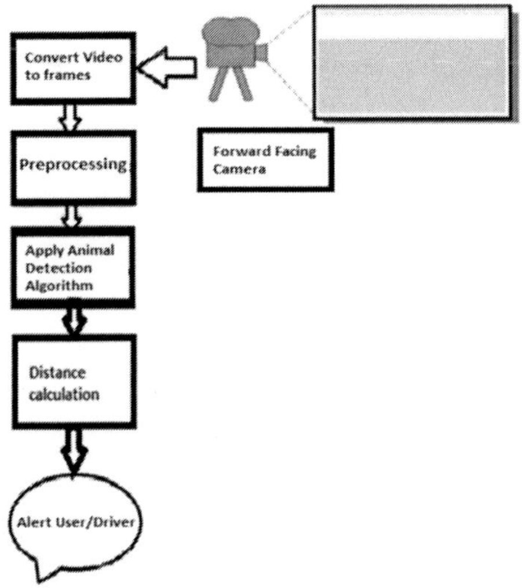

Fig. 12: The suggested Animal Detection and Collision Avoidance System's architecture

HOG Algorithm:

The HOG Algorithm consists of 5 steps, each of which are discussed in detail below as in Fig. 13, Fig. 14:

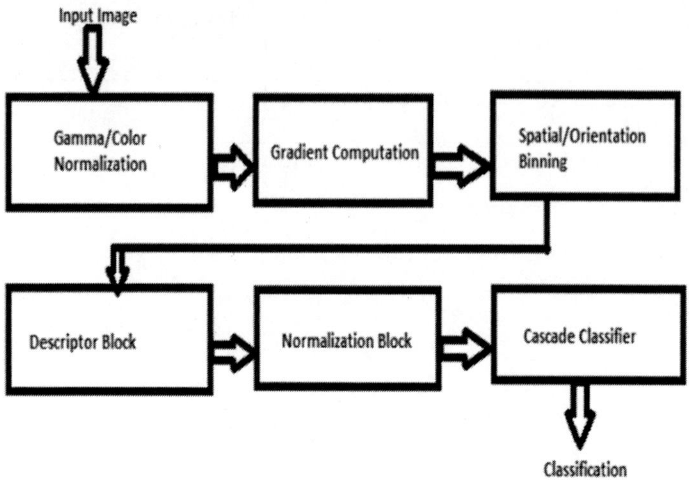

Fig. 13: HOG Algorithm

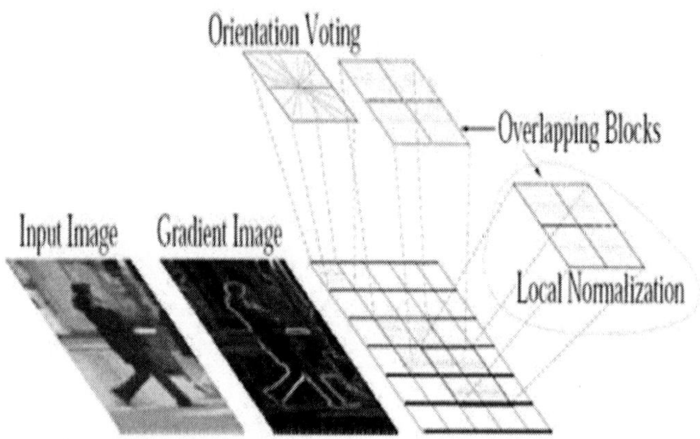

Fig. 14: Block Normalization scheme

Step 1: Colour Normalisation

The input image is first supplied to the colour normalisation block, which is used to remove intensity values while maintaining the integrity of the colour values while detecting objects on colour images. The gradient values must then be calculated.

Step 2: Computation of Gradient Values

One of the methods that is widely used is to apply the 1D centred point discrete derivative mask in both the horizontal and vertical axes. The following kernels must first be used to filter the grayscale image before doing this:

$$D_X = [-1 \quad 0 \quad 1] \quad \text{and} \quad D_Y = \begin{bmatrix} 1 \\ 0 \\ -1 \end{bmatrix}$$

Hence, for a given input image ‘I’, the x and y derivatives can be calculated using a convolution operation:

$I_X = I^* D_X$ and $I_Y = I^* D_Y$

The magnitude of the gradient, $|G| = \sqrt{I_x^2 + I_y^2}$

The orientation of the gradient, $\theta = \tanh^{-1}\left(\frac{I_y}{I_x}\right)$

Step 3 - Creating Cell Histograms

A rectangle cell has many pixels contained within it. Based on the gradient values generated for each pixel in the previous phase, a weighted vote is made (for each pixel) for an orientation-based histogram channel. Depending on whether the gradient is "signed" or "unsigned," the histogram channels are equally distributed over 0 to 180 degrees or 0 to 360 degrees, respectively.

The pixel contribution vote weight is equal to the gradient magnitude divided by the gradient magnitude's square root. To eliminate the impacts of fluctuations in light and contrast, the gradient strengths must be locally adjusted.

Step 4: Normalisation of Gradient Strengths

Cells must be arranged into bigger, spatially related blocks in order to normalise gradient strengths. The HOG descriptor is then made up of this vector of normalised cell histogram components (from all of the block sections). These blocks regularly cross over, indicating that each cell contributes more than once to the overall description.

Blocks can come in two main geometries:

- **Rectangular R-HOG blocks:** Three parameters are used to represent the square grids known as R-HOG blocks. They are: the number of cells per block, the number of pixels per cell, and the number of channels per cell histogram.
- **Circular C-HOG blocks**

Among the different methods of block normalisation, the following method is followed:

Assumptions:

Let,

v = non-normalized vector containing all histograms in a given block

$||v_k||$ = its k-norm for k = 1, 2

e = a small constant (this value will not have a major impact on the results)

Then, one of the following can be the normalization factor

$$\text{L2-norm:} f = \frac{v}{\sqrt{\|v\|_2^2 + e^2}}$$

$$\text{L1-norm:} f = \frac{v}{\|v\|_1 + e}$$

$$\text{L1-sqrt:} f = \sqrt{\frac{v}{\|v\|_1 + e}}$$

Step 5: Classification of Image:

The image would then finally go to the cascade classifier for the classification of the object within the image. Compared to other descriptors, the HOG descriptor is best suited for detecting animals in videos or photos because of the following benefits:

- It is unaffected by geometric and photometric modifications because it works with local cells.
- If animals retain a nearly upright posture, coarse (spatial) sampling, strong local photometric normalisation, and fine orientation sampling enable distinct body movements to be disregarded.

Cascading:

An instance of group-based learning is cascading, which is the concatenation (a series of additions) of different classifiers. This entails providing the next classifier in the group with all the data gathered from the output of the previous classifier [24]. Boosted cascade classifiers [25], as shown in Fig. 15, are found to be more advantageous than monolithic classifiers owing to their ability to learn faster and use a low computation time.

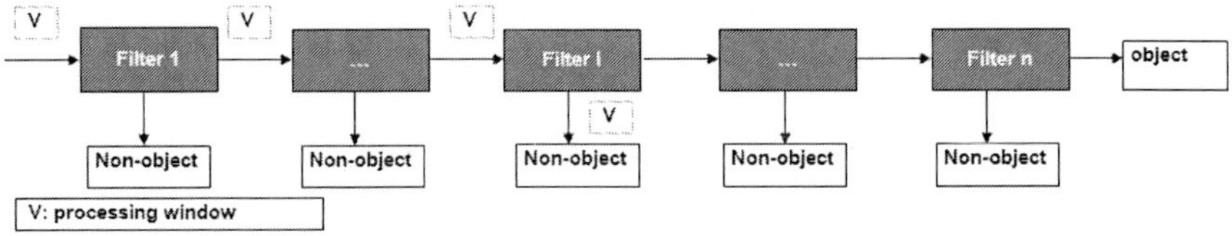

Fig. 15: Boosted Cascade Classifier

As illustrated in Fig. 15, each filter in the boosted cascade classifier rejects non-object windows while allowing object windows to pass through and advance to the next layer of the cascade. A window is regarded as an object if and only if all tiers of the cascade classify it as such. The filters in the cascade are built to:

- Discard the potentially numerous "non-object" windows.
- Allow rapid inspection of the possibly numerous object windows.

Hence, the cascading process eliminates the candidates (false positives) early on, owing to its improved efficiency.

Dataset used: KTH dataset [26] and NEC dataset [27]

The dataset consists of additional images clicked during different times and weather conditions and in total has 2200 images – positive and negative images to give robustness. Upon developing the classifier, it was tested with various videos taken from the camera with a frame rate of 30 fps mounted on the vehicle.

Training and Testing:

Considering the animal distribution in India, the most common animals found here, are cows, dogs, cats, and oxen. India consists of more than 20 varieties of cow. The images of all these varieties of cow have been added to the dataset. The following procedure was followed:

- Positive (Fig. 16) and Negative (Fig. 17) images have been collected into a data folder.
- Annotation had been generated.
- A sample was created—a generate.vec file.
- The data had been trained, and an XML file had been generated. The training parameters are as shown in Fig. 18.
- Testing was done.

Fig. 16: Positive images Fig. 17: Negative images

Parameters	Value/Type
numPos (number of positive samples)	700
numNeg (number of negative samples)	1500
numStages (number of stages in cascade)	20
stageType (type of stage in cascade)	BOOST
featureType (feature type for extraction)	HOG
sampleWidth (width)	70 pixels
sampleHeight (height)	40 pixels
boostType (type of boosting)	GAB (Gentle AdaBoost)
minHitRate (minimum hit rate of the classifier)	0.995
minFalseAlarmRate (minimum false alarm rate of the classifier)	0.5

Fig. 18: Parameters utilised during the model training

Calculation of Distance:

The captured video is converted into image frames of size 640 × 480. Hence,

Image resolution = 640 × 480

X range = 0 to 640

Y range = 0 to 480

Now, assuming a flat surface, the distance between the animal and the lower edge camera (connected to the car) is 480-y if the bottom-right coordinate of the identified animal is (x, y) (in case of non-flat surfaces, the value will differ).

Calculation of depth of the object in meters:

It has been found that the relationship between an item's distance from a camera-mounted vehicle in pixels and its distance from the object in metres is exactly proportionate; as one value rises, the other rises as well [7]. This allows us to determine the link between an object's depth in pixels and its depth in metres.

In order to achieve this, the position of the camera mounted on the car and its height above the ground should be fixed (camera calibration is done). After taking multiple pictures of the same object held at different distances from the camera's centre, it was determined that the object's depth in metres was a "known value." The associated pixel values were located and tabulated as depicted in Fig. 19.

Depth (mt)	1	2	3	4	5	6	7	8	9	10	11	12	13	14
Depth (pixel)	23	45	69	91	114	136	159	180	206	226	245	274	295	320

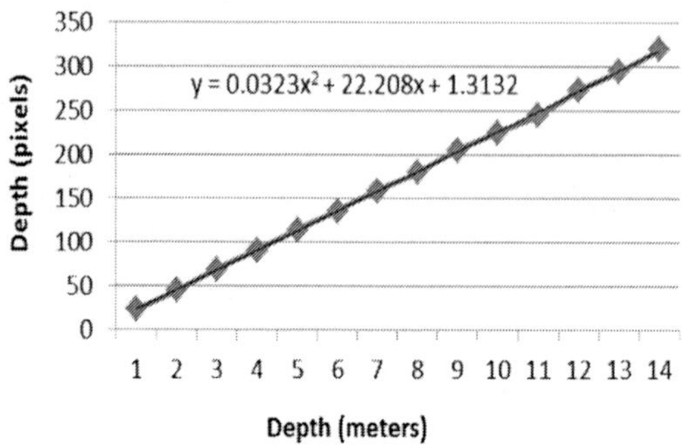

Fig. 19: A plot of the relationship between pixels and meters.

Based on the above study results, the following polynomial equation is formed:

$$y = 0.0323x^2 + 22.208x + 1.3132$$

where,

y = depth in pixels

x = depth in meters

This formulation was tested, the actual and observed values were compared and it was found that the error was less than 2 percent.

Observation:

Object detection in computer vision and image processing is possible using feature descriptors called HOG descriptors [28].

- ***System configuration:*** Intel(R) Core(TM) i5-2430M CPU, 2.40GHz, 4GB RAM.
- ***Average time taken for the training and testing:*** 14 hours [specific to system configuration listed above].

- ***Hardware configuration (camera):*** ASUS x53s, Intel(R) Core(TM) i5-2430M CPU, 2.40GHz, 4GB RAM
- ***Software used:*** Microsoft Visual Studio 10 Professional, OpenCV 2.4.3, 64-bit operating system running under Windows 7.

The parameters required for testing the developed classifier are:

- Sensitivity [or the True positive rate] = $\frac{TP}{(TP+FN)}$
- Specificity [or the True negative rate] = $\frac{TN}{(TN+FP)}$
- Accuracy = $\frac{(TN+TP)}{(TN+TP+FN+FP)}$

where,

TN = true negative, i.e., the animal is not detected when it is not there, as shown in figure 20.

TP = true positive, i.e., an animal is detected when the animal is there.

FN = false negative, i.e., despite being present, the animal is not noticed. as shown in figure 21

FP = false positive, i.e., detection of an animal when none is present, as shown in Figure 22.

TP and TN are the most pertinent and accurate categorization parameters out of the four (TN, TP, FN, FP).

Fig. 20: True positive case; Fig. 21: False positive case; Fig. 22: False negative case

Result:

Out of a total of 640 frames, 105 frames show animal detected even when there is no animal present. Hence,

FP = 105

TN = 535

125 frames out of a total of 640 show no animal detected even though an animal is there.

Hence,

FN = 125

TP = 515

From this, the following results are calculated:

- Sensitivity = 80.4%
- Specificity = 83.5%
- Accuracy of the classifier = 82.5%.

A. Blockchain related approach and use case

[31] Blockchain is a decentralised, unchangeable database that streamlines asset tracking and transaction recording inside a business network. An asset may be tangible (such as a house, a car, cash, or a piece of land) or intangible (intellectual property, patents, copyrights, and branding). Practically everything of value may be recorded and traded on a blockchain network, reducing risk and boosting efficiency for all parties.

[30] Critical concerns to be solved with respect to blockchain environment are data privacy and system dependability. An anomalous intrusion can seriously endanger a blockchain. A study conducted regarding the same suggests a cooperative data fusion strategy based on clustering characteristics for intrusion detection in a blockchain-based system.

Here, an AI model and a mathematical model of data fusion are used to train and evaluate the data clusters in blockchain networks. After locating the anomalous features in a blockchain dataset, conducting a weighted combination, and accumulating the weighted coefficients among several clustering nodes following numerous iterations of mutual competition between clustering nodes

In conclusion, a blockchain provides us with decentralised access. Access restriction provides protection and trust in the chain. A use case of the said explanation could be the detection of animals during safaris owned by a particular sanctuary. For the sanctuary to preserve their data without any intrusion or leaks, a blockchain technique could be applied to the stored animal detection algorithm, thus preventing any loss for the sanctuary in terms of valuable data fetched and owned by them.

3. Conclusion

This paper gives a detailed study of some of the popular approaches over the past 22 years. From the approaches discussed so far, the most effective and trending approaches are the modern approaches. Among the four modern approaches discussed, the deep CNN-based solution and the HOG and cascade classifier can be concluded to be more effective due to the accuracy they provide and the experimental results obtained in the referenced papers. Additionally, Block chain security techniques can be applied on top of the fetched data to make collected data confidential and protected.

References

[1] Kakkar, R., Aggarwal, P., Kakkar, M., Deshpande, K., & Gupta, D. (2014). Road traffic accident: retrospective study. *Indian Journal of Scientific Research*, 5(1), 59.

[2] Government of India - Ministry of Road Transport and Highways - Transport Research Wing 2019). *Road Accidents in India 2019*. Ministry of Road Transport and Highways. Retrieved June 11, 2022, https://morth.nic.in/sites/default/files /RA_Uploading.pdf

[3] Saxena, A., Gupta, D. K., & Singh, S. (2021). An animal detection and collision avoidance system using deep learning. *In Advances in Communication and Computational Technology* (pp. 1069-1084). Springer, Singapore.

[4] Chakraborty, P. (2021, August 12). *Nine stray animals injured in road accidents in Nagpur every day over 52 months*. The Times of India. Retrieved June 6, 2022, https://timesofindia.indiatimes.com/city/nagpur/9-stray-animals-injured-in-road-accidents-in-city-every-day-over-52-months/articleshow/85253789.cms

[5] Mohanty, C. R., Radhakrishnan, R. V., Jain, M., Sasmal, P. K., Hansda, U., Vuppala, S. K., &Doki, S. K. (2021). A study of the pattern of injuries sustained from road traffic accidents caused by impact with stray animals. *Journal of Emergencies, Trauma, and Shock*, 14(1), 23.

[6] *Cattle the real menace in Indian highways. . ..* (2013, November 10). Team-Bhp.Com. Retrieved June 9, 2022, https://www.team-bhp.com/forum/road-safety/144117-cattle-real-menace-indian-highways.html

[7] Sharma, S. U., & Shah, D. J. (2016). *A practical animal detection and collision avoidance system using computer vision technique*. IEEE Access, 5, 347-358.

[8] Kakkar, R., Aggarwal, P., Kakkar, M., Deshpande, K., & Gupta, D. (2014). Road traffic accident: retrospective study. *Indian Journal of Scientific Research*, 5(1), 59.

[9] Burghardt, T., &Ćalić, J. (2006). Analysing animal behaviour in wildlife videos using face detection and tracking. *IEE Proceedings-Vision, Image and Signal Processing*, 153(3), 305-312.

[10] P. Viola and M. Jones. Robust real-time object detection. *Second International Workshop on Statistical and Computational Theories of Vision*, 2001.

[11] T Troscianko, C A Parraga, P G Lovell, D Tolhurst, R Baddeley, and U Leonards. Natural illumination, shadows, and primate colour vision. ECVP European Conference on Visual Perception, 2004.

[12] Nascimento, J. C., & Marques, J. S. (2006). Performance evaluation of object detection algorithms for video surveillance. *IEEE Transactions on Multimedia*, 8(4), 761-774.

[13] N. Kawasaki, ''Parametric study of thermal and chemical non equilibrium nozzle flow,'' M.S. thesis, Dept. Electron. Eng., Osaka Univ., Osaka, Japan, 1993.

[14] S. L. Hannuna, N. W. Campbell, and D. P. Gibson, ''Identifying quadruped gait in wildlife video,'' in *Proc. IEEE Int. Conf. Image Process*. (ICIP), Genoa, Italy, Sep. 2005, pp. 713–716.

[15] D. P. Gibson, N. W. Campbell, and B. T. Thomas, ''Quadruped gait analysis using sparse motion information,'' in *Proc. Int. Conf. Image Process*. (ICIP), New York, NY, USA, Sep. 2003, pp. 333–336.

[16] Ragheb, H., Velastin, S., Remagnino, P., & Ellis, T. (2008, October). Human action recognition using robust power spectrum features. In *2008 15th IEEE International Conference on Image Processing* (pp. 753-756). IEEE.

[17] Wichmann, F. A., Drewes, J., Rosas, P., & Gegenfurtner, K. R. (2010). Animal detection in natural scenes: Critical features revisited. *Journal of Vision, 10*(4), 6-6.

[18] P. Viola and M. J. Jones, ''Robust real-time face detection,'' *Int. J. Comput. Vis*., vol. 57, no. 2, pp. 137–154, 2004

[19] D. Ramanan, D. A. Forsyth, and K. Barnard, ''Building models of animals from video,'' *IEEE Trans. Pattern Anal. Mach. Intell*., vol. 28, no. 8, pp. 1319–1334, Aug. 2006.

[20] M. S. Zahrani, K. Ragab, and A. U. Haque, ''Design of GPS-based system to avoid camel-vehicle collisions: A review,'' *Asian J. Appl. Sci.*, vol. 4, no. 4, pp. 362–377, 2011.
[21] S. Shaikh, M. Jadhav, N. Nehe, and U. Verma, ''Automatic animal detection and warning system,'' *Int. J. Adv. Found. Res. Comput.*, vol. 2, pp. 405–410, Jan. 2015.
[22] Mitra, V., Wang, C. J., & Edwards, G. (2003, July). Neural network for LIDAR detection of fish. In *Proceedings of the International Joint Conference on Neural Networks*, 2003. (Vol. 2, pp. 1001-1006). IEEE.
[23] Banupriya, N., Saranya, S., Swaminathan, R., Harikumar, S., & Palanisamy, S. (2020). Animal detection using deep learning algorithm. *J. Crit. Rev*, *7*(1), 434-439.
[24] P. Viola and M. Jones, ''Rapid object detection using a boosted cascade of simple features,'' in *Proc. Comput. Vis. Pattern Recognit.*, Dec. 2001, pp. I-511–I-518.
[25] Jones, C. (2015). *Cascade Classifier*. Slideplayer. Retrieved June 9, 2022, from https://slideplayer.com/slide/5754730/19/images/49/Cascade+Classifier+Attentional+cascade+False+Positive+Rate%3A.jpg
[26] KTH *Animal Datasets*, accessed on Jun. 18, 2013. [Online]. Available: http://www.csc.kth.se/~heydarma/Datasets.html
[27] NEC *Animal Datasets*, accessed on Jun. 18, 2013. [Online]. Available: http://ml.nec-labs.com/download/data/videoembed
[28] N. Dalal and B. Triggs, ''Histograms of oriented gradients for human detection,'' in *Proc. IEEE Comput. Soc. Conf. Comput. Vis. Pattern Recognit.* (CVPR), New York, NY, USA, Jun. 2005, pp. 63–69.
[29] Keel, M., &Trentz, O. (2005). Pathophysiology of polytrauma. *Injury*, 36(6), 691-709.
[30] Liang, W., Xiao, L., Zhang, K., Tang, M., He, D., & Li, K. C. (2021). Data fusion approach for collaborative anomaly intrusion detection in blockchain-based systems. *IEEE Internet of Things Journal*.
[31] *What is Blockchain Technology? - IBM Blockchain | IBM. (n.d.).* What Is Blockchain Technology? - IBM Blockchain | IBM. Retrieved January 10, 2023, from https://www.ibm.com/in-en/topics/what-is-blockchain

CHAPTER 4

MANAGING ACCESS CONTROL AND AUTHORIZATION SCHEMES OF HUGE ENTERPRISE RESOURCES USING MACHINE LEARNING (ML) AND ARTIFICIAL INTELLIGENCE (AI) TECHNOLOGIES

MICHAEL OLAJIDE, ADEKOLA AJAYI, OLUWAGBOYEGA AFOLABI, OLADOYIN ABIODUN

ADEYEMI FEDERAL UNIVERSITY OF EDUCATION, ONDO, NIGERIA

Abstract

A decentralised type of data that is connected over a single shareable resource is made available by blockchain technology. The problem of access control and authorization becomes crucial to take into account because blockchain technology introduces a distributed solution that does not involve a third-party transaction. This work intends to consider the design of a broad framework that can conveniently handle access control and authorization schemes for integrated enterprise resources using machine learning (ML) techniques and artificial intelligence (AI) technologies to guard against security breaches. If it is implemented, the benefits of the new framework can be identified and explained to provide a future direction for further work.

Keywords: access control, authorization, machine learning, artificial intelligence

Introduction

A blockchain is a straightforward list of connected blocks, where an individual block includes multiple transactions as well as multiple pieces of information, as stated by Messika on August 29th, 2018. The concept of blockchain makes provision for a distributed, immutable form of data that is utilised by a collection of people that are directly connected across a common shareable resource, often referred to as a database, and ultimately creates electronic assets. The transactions are usually recorded and can be queried from time to time in an environment of trust and transparency.

According to Memon, Syed, Umair, and Asad (2018), the objective of blockchain technology is to create a decentralised solution that does not necessitate a third-party transaction. They posited that blockchain, over time, has demonstrated itself to be a very rich solution to many issues, which may include

1. Exchange of resources (data) among several organisations
2. Digital asset registration (coins or bitcoin, for instance)
3. Integrity and identity management as obtained in cryptography

In this work, attention will be drawn to the first issue (a), where access control and authorization schemes for huge enterprise resources that are fully inter-organisational will be accorded prominence using machine learning techniques as the front-end and artificial intelligence techniques as the back-end.

Literature Review

The ever-increasing dependence on enterprise data and resources for the driving of its activities is a global phenomenon. Data and electronic systems made this possible. The premium placed on enterprise resources as a result of strong momentum has further put pressure on some enterprises with similar or almost entirely similar resources to forge a good measure of integration (Seethamraju 2015, 479). The main goal is to enable the integration to provide sufficient infrastructure with very robust features and services. One serious and vital strength is the preservation of the enterprises' resources, which is a priceless asset of such organisations. On a daily basis, there are cases of breaches of privacy and privileges in cyberspace everywhere. The interlopers, hackers, and other notorious electronic hawks perform no other task than compromising identities within cyberspace, leaving behind a massive and devastating economic loss. Therefore, there is a compelling reason for giving concessional consideration to the aspect of enterprises' access control and authorization of their corporate resources. The pride and strength of the success of enterprises having integrated resources hosted within cyberspace is the ability to meticulously enforce strong security cover for this asset (Wallace and Kremzar 2002, 12–15). In addition, it takes a step further by providing other pre-emptive necessary steps, which can further bolster the existing security measures as an additional external security layer for the common resources (asset). Access control simply determines who does what, when to do it, when not to do it, and to what depth, based on a well-designed policy. In most cases, the heartache for administrators is controlling valuable enterprise resources (assets). One prominent part of an organisation's activities is the access control mechanism placed on their resources (Jamieson and Handzic 2004, 481-482). This can be direct security issues bothering major activities in the organisation, as well as designing appropriate security requirements that also consider periodic audits and evaluations of enterprise resources and performance. The increasing expansion of cyberspace has necessitated pronounced authorization schemes that are detailed enough to enhance the protection of enterprise resources. Access control and authorization are key ingredients in the existence of electronic enterprises, especially those that manage robust common resources for similar organisations.

Authorization is the aspect of security involving checking the user's identity to confirm permission status to access any portion of the resource provided the user successfully gets authenticated (Joshi, Shalini, Prashant, Piyus, et al., 2021, 3–4). Authorization is largely credential-focused and adheres to established standards and access control lists that web application administrators have previously established. Typical permission checks entail looking for a user's clearance, user group membership, or presence on an access control list that has been approved by the resources department (an electronic bouncer). Effective and forged-resistant authentication rules used for permission are necessary for every system (Mohammed 2012, 3–6). On the other hand, access control refers to "a broader method of restricting access to web resources, which may include limitations based on the time of day, the IP address of the HTTP client browser, the type of encryption the HTTP client can support, the number of times the user has authenticated that day, the possession of any number of different types of hardware or software tokens, or any other derivable indices that can be easily calculated" (Gusmeroli, Salvatore, and Domenico2013, 1192–1193). The basic access control element consists of the following: subjects, objects, and access rights. The entity that has access to objects is the subject. The

The term "object" refers to the asset to which access is restricted, while the term "access right" refers to how a person may access the object. The access right will ultimately require authentication and authorization key issues that confer due access rights and accord privileges to the subjects. The topic simply attempts to explore access control and authorization schemes in an integrated enterprise resource within cyberspace with the infusion of machine learning (ML) and artificial intelligence (AI) techniques. A design of a quality framework that is capable of driving several enterprises' resources together for ultimate utilisation for the attainment of enterprises' corporate objectives.

Motivation

It is a known fact that enterprise resources are a strong factor of production in the emerging knowledge society. Most enterprises that are electronically driven will always prefer to store or keep their assets in electronic archives in cyberspace. Nowadays, the dependence of the global enterprise on electronic processing and storage of resources will continue to soar as the benefits immensely improve. The malicious activities of hackers and interlopers on global networks have also increased, leaving behind distasteful experiences for unsuspected victims. The experience of most enterprises has not been captivating, considering the marauding effects of the negative impacts bequeathed to this network by a group of people whose interest is to compromise and steal valuable assets belonging to these legitimate business organisations. The growing population of the knowledge society has profound instincts to get most of their daily activities arranged and organised by the emerging electronic technology, leaving almost everything that could be done manually with the technology.

Also paramount is the fact that most designs and devices released for the consumption of members of the emerging knowledge society are strongly tied to using the Internet and its associated resources to control and manage them to meet their immediate and routine needs. It is on the strength of this, therefore, that individual enterprises base their operational activities and performance on electronic technology. In addition, several enterprises nowadays combine to facilitate inter-organisational relationships by having corporate assets shared as the main goal. It is always good to have a group of similar enterprises fuse and utilise a robust and common databank or asset for the effective pursuit of their different or similar corporate objectives. The concept of inter-organisational fusion of resources is generally seen as cost-effective and relatively economical.

This work intends to consider the design of a broad framework that can conveniently handle access control and authorization schemes for integrated enterprise resources using machine learning (ML) techniques and artificial intelligence (AI) technologies.

Objectives

1. Present access control and authorization schemes as a vital tool in the enterprise resource domain.
2. Appraise machine learning techniques.
3. Appraise artificial intelligence technologies.
4. Design a framework for robust access control and authorization schemes in a domain that is driven by ML techniques and AI technologies.

Methodology (the proposed content of the work)

The methodology of this work begins with the introductory section, followed by the section that appraises the existing literature. Thereafter, a framework that presents a broad-based integrated enterprise resource model that is pivoted by ML techniques and powered by AI technologies will be considered. Finally, the benefits of the new framework, if implemented, can be identified as a future direction for further work.

Fundamental concept of artificial intelligence

Since John McCarthy first used the term "artificial intelligence" in 1956, when the idea first gained attention following World War II, several developments have occurred, with innovations springing up as a result of embracing and employing AI. Almost all industries, including the automotive, entertainment, healthcare, marketing, and science and technology sectors, now consider AI a necessary topic. Due to their adaptability and myriad advantages, the majority of them have been incorporated into their operations. Due to significant investment, this AI technology has become relevant in many industries and is transforming how people live their lives. Similar assertions claim that artificial intelligence (AI) is a vital technology that underpins daily economic and social activities and that it is becoming more prevalent in the management science and operational research fields of study (Pannu 2015, 79–80). These are the advantages that society experiences as a result of the creation and evolution of AI. Commonly, intelligence is defined as the capacity to acquire knowledge and use the knowledge acquired to reason about situations, both basic and complex. Since AI has improved human lives in so many ways, it is becoming more and more popular in computer science. AI has significantly increased the efficiency of industrial and service systems over the last 20 years (Lu, Yujie, Min, and Hyoungsepet al. 2018, 368–370).

AI has been defined and described in a variety of ways by different academics and authorities. "Machines are built to think like people and will even think better than humans in years to come," is the general tenet of AI. According to Singh and Sagar (2013, 59–63), "AI is the area of computer science that focuses on researching and developing intelligent systems that can perceive their surroundings and take actions that will enhance their chances of success". According to Saleh (2019, 871), AI explains the form of intelligence in which "machines display intelligence to carry out a variety of activities such as speech recognition, learning, planning, and problem-solving." A further definition of AI is "the study and creation of intelligent machines and software that can reason, learn, gain knowledge, communicate, manipulate, and sense objects" (Pannu 2015, 79–80). AI is also a powerful tool that enables machines to learn from their experience, adjust to new changes, and carry out activities identical to those performed by humans (Wamba-Taguimdje, Samuel, Jean, and Chris et al., 2020, 1899–1901).

On occasion, AI might be thought of as the study of the computations that enable perception, reasoning, and action. Because it has a strong emphasis on computation, it cannot be compared to psychology. It also differs from computer science in that it places a strong emphasis on perception, thinking, and action. Machines become smarter and more useful overall, thanks to AI, and this usefulness is valued in almost every aspect of life. AI is also distinct from natural intelligence and has advantages over the latter because it can be documented, is more permanent and consistent, is less expensive, and can execute some activities more quickly and proficiently than humans

Components of Artificial Intelligence

Vision (also known as computer vision), expert systems, heuristic problem-solving, and natural language processing are the four main components of AI. They are discussed briefly below.

Vision: Automatic object and feature recognition is a capability of machines. Computers and other systems can use computer vision to gather useful data from digital photos, movies, and other visual inputs and act on or make suggestions for such data (Bekhit and Ahmed 2022, 15–16). Computer vision provides robots with the ability to detect, observe, and comprehend, just like AI affords them the facility to think. Computer vision is used in the energy, utility, manufacturing, and automobile industries, and the market is continually growing. Computer vision requires a lot of data. It does data analysis frequently to draw distinctions and eventually identify images (Umbaugh 2010, 65–69). For example, to teach a computer to recognise vehicle tyres, a massive amount of tyre pictures and tyre-related items must be provided to it. This is especially true of flawless tyres.

Expert Systems: An expert system is a piece of technology that memorises relevant human experience to offer sage advice and support demands or decisions. Expert systems rely on a large, properly organised store of specialised information on a particular topic (Singh and Sagar 2013, 59–63). Expert systems are useful because they have been shown to perform better than people, which results in fewer errors and more consistency in their suggestions. Expert systems can facilitate operations in hostile environments and generally increase industrial output.

Heuristic problem solving: This method of problem solving produces an immediate and workable solution. Heuristics can be utilised to provide poor but acceptable results in circumstances where flawless answers may be improbable. The majority of heuristic approaches involve making decisions based on past experiences by applying mental shortcuts. For instance, if a student had to decide what to major in at university, he or she would probably trust his or her gut and pick the subject matter that interests, excites, and rewards them the most. Heuristic methods are also used by financial experts and investors to expedite analysis and investment choices.

Natural Language Processing (NLP): This field combines linguistics, artificial intelligence, and computer science to give robots the ability to understand spoken or written words in human language (voice data) like that of people (Priya, Nandhini, and Gnanasekaran 2021, 600). Using artificial intelligence, NLP processes and converts any language, whether spoken or written, so that the computer can understand it. Similar to how humans have ears to hear and eyes to see, computers read from programmes and capture audio using microphones. Additionally, computers process input using programmes and algorithms tailored to the appropriate inputs, just like humans do with their brains. The input is ultimately transformed into a type of code that the computer can comprehend.

Approaches to the general functionality of artificial intelligence

By utilising algorithms to comprehend human goals or strategies for achieving those goals, AI mimics human intelligence. For a better understanding of the objective, it develops a relationship between goal-seeking, data processing, and acquisition. In light of it, the following list contains four AI approaches (Lieto, Mehul, Alessandro, and David 2018).

1. **Acting Humanly:** It can be difficult to distinguish between a computer and a human when using technologies like automated reasoning, machine learning, and NLP. Without any direct physical interaction, the Turing test, also known as an imitation game, determines whether or not a machine can demonstrate human intelligence or consciousness.
2. **Thinking Humanly:** When a computer can execute functions that require human intellect, such as driving a car, and can think just like a person, it becomes imperative to conclude that the machine is thinking. To determine how people think, the cognitive modelling approach, which is based on three methods—introspection, psychological testing, and brain imaging—is utilised. This category of the human mind is used to create precise models in the field of psychology as well as in the health sector when appropriate.
3. **Thinking Rationally:** Some principles that support creating a moral code for individuals are used in the conventional study of the human mind. The computer solves problems in the same way a person who is considered to be rational would think. Because computers rely on such logical processes to operate, the answer to a given problem may not be the same as the actual solution.
4. **Acting Rationally:** Rational beings are necessary for the study of human conduct in complex or ambiguous Actions are based on the

situation, the surrounding elements, and the information at hand to maximise the expected value of their performance. To get the intended outcome, black-box or engineering approaches are frequently used.

Characteristics and Features of Artificial Intelligence

The following are a few of the elements that need to be studied to have a thorough understanding of AI because it has changed since its conception and to justify the hype that has been created around this technology (Tekerek and Bay, 2019, 190–193).

1. **Intelligent Robotics:** Robotics combines engineering, science, and technology to produce programmable robots that can assist humans or mimic their creators. Robots were primarily developed to perform repetitive tasks, but they are now employed in a variety of domestic, business, and even military settings. Robots of today range from human-controlled to autonomous, each with a different degree of autonomy to do tasks independently.
2. **Perception:** Processing and making inferences from all the data included in sensor inputs is feasible with the help of machine perception. Object, facial, and speech recognition are some of the applications that predominantly use it. Only computer vision is available for analysing visuals.
3. **Imitation of Human Cognitive Process:** Artificial intelligence is capable of imitating human intelligence and can reply to some straightforward client questions with audio or text. Artificial intelligence-enabled computers called chatbots provide a window to hear about customer issues and provide real-time responses to requests (Laskar, Abdullah, Partha, and Rabiah et al. 2022, 123). These intelligently created bots can quickly connect you to human administrators and executives if they are unable to resolve your issue within a set of criteria.
4. **Quantum Computing:** Complex quantum physics calculations are another area that AI has dominated. You may obtain correct answers from supercomputers using quantum neural networks, and the time will soon come when artificial intelligence will assist us in making ground-breaking discoveries. The foundation of the interdisciplinary discipline of quantum computing, which advances computational tasks, is creating extraordinarily complex quantum

algorithms (Schuld and Petruccione 2018, 65). The entire concept was developed using quantum-enhanced AI.

5. **Cloud Computing:** One of the most common applications of artificial intelligence is cloud computing. As huge volumes of data are produced every day and physical storage is becoming a bigger problem, cloud computing has offered a better way. The advent of cloud computing is the best solution for organisations to adapt to and make them more efficient, observant, and strategic (Avram 2014, 530).

Types of Artificial Intelligence

Based on functionality and capabilities, AI can be categorised into two categories: These are listed below.

1. Reactive Machines: These are the earliest and most fundamental types of artificial intelligence (AI) systems. These tools have very few uses. Since these systems are unable to remember the past or use it to inform the present, they cannot "learn" and cannot get better with practise. They can only perceive and respond to the world around them using their intelligence. A reactive machine cannot utilise past experiences to inform current decisions since it lacks memory. One well-known example of a reactive machine is "Deep Blue," which was developed at This machine was deemed perfect and effective as it was able to defeat a chess grandmaster in 1997 (Kelly and Long 2010, 60).
2. Limited Memory: This kind of AI accumulates experiential knowledge by keeping an eye on actions or data and studying the past. Machine learning models with a finite memory are used to discover insights from previous facts, events, or information that has been retained. To create predictions and manage challenging categorization tasks, this kind of AI combines pre-programmed information with historical, observational It is now the most commonly used type of AI. Self-driving automobiles are examples of limited memory AI which leverage information gathered recently to make swift choices. For example, self-driving cars use sensors to detect people crossing the street, sloping roadways, traffic lights, and other hazards to help them make better driving decisions (Hecht 2018, 30).
3. Theory of Mind: This level of cognitive comprehension entails the capacity to consider the abstract mental states of both our own and

others. A requirement for integrating artificial intelligence (AI) into human society is the ability to discern human mental states. Theory of mind AI constructs have the potential to have a substantial impact on empathic healthcare, for example, in interactions with people who are suffering from neurological diseases such as Alzheimer's as well as mental conditions like depression, autism spectrum disorder, and schizophrenia (Cuzzolin, Alice, Bodgan, and Barbara 2020, 1059). These methods may increase the effectiveness of psychological therapies like cognitive behavioural therapy or mindfulness.

4. Self-Aware AI: The machines in this sort of AI are aware of their internal states as well as those of others and their emotions, behaviours, and intelligence (Picard and Klein 2002, 159). If this AI materialises in the future, we will surely witness a robot with consciousness and intelligence on par with.

Following are the types of AI based on capabilities:

1. Narrow Artificial Intelligence (NAI): This particular kind of artificial intelligence creates a learning algorithm to finish a single task; any insights acquired from finishing that work won't automatically be extended to subsequent assignments. NAI is made to finish a single task successfully without help from a human. Language translation and image recognition are two common uses for NAI (Goralski and Tan 2020, 82).
2. Artificial General Intelligence (AGI): This term, which is also used to refer to general AI, refers to a particular type of AI that, rather than focusing on more specialised or narrower types of tasks, is broad in the same sense that cognitive processing systems can handle a wide range of tasks effectively and accurately, simulating the depth of the human intellect (Kuusi and Heinonen 2022, 67).
3. Artificial Super Intelligence (ASI): This artificial intelligence (AI) system is capable of self-awareness and intelligence levels that are superior to those of humans (Gill 2016, 139). According to certain views, superintelligence, or artificial intelligence, will soon surpass human intelligence in the majority of its aspects (Pueyo 2018, 1735).

Application Areas and Innovations in Artificial Intelligence

Artificial intelligence is used in a variety of domains of endeavour. These range from e-commerce (for individualised purchasing, artificial intelligence-powered virtual assistants for shopping, and fraud detection) to education (teaching assistants, individualised learning, and so forth). Numerous ground-breaking inventions have come up recently as a result of AI research. Robots that are capable of carrying out translation activities in real-time have been made public by Microsoft Corporation, along with other cutting-edge picture recognition technology. A cleaning robot and a four-foot walking robot have been created at the Massachusetts Institute of Technology's Computer Science and Artificial Intelligence Laboratory. DeepFace, a facial recognition system created by Facebook, is based on artificial intelligence (Taigman, Ming, Marc'Aurelio, and Lior 2014, 1705).

Applying Artificial Intelligence to Security

Organisations will be able to improve their current application security skills with the help of AI in cyber security. Application security safeguards applications that go through multiple stages of the security development life cycle, such as security design and secure coding, such as mobile, web, or thick client apps (Maraju 2018, 4). Application security can be enhanced by utilising a variety of AI disciplines, such as machine learning and expert systems, to produce, anticipate, or apply inferences to forecast security risks, identify security vulnerabilities, and identify security code rectification recommendations (Anwar and Hassan 2017, 884). Decision Tree Learning (DTL) for machine learning helps identify threats, while forward and backward chaining for expert systems is helpful for security code review and code review advice. DTL uses a training set of examples to construct a hypothesis and derive generalisations. It also selects the most advantageous assault paths when it can (threats). Expert systems, on the other hand, can evaluate the input made, offer recommendations, and draw a conclusion. They make predictions about the results and suggest potential fixes.

Artificial Intelligence in Network Intrusion Detection

To defend computer and communication networks from intruders, intrusion detection systems (IDS) employ a variety of AI techniques. One of the AI tools is the Artificial Neural Network (ANN), which is a mathematical model made up of a network of artificial neurons that process information and are helpful for IDS. ANN is used to detect patterns in data or to simulate complex interactions between inputs and outputs. As a result, it is simple and rapid to find unlawful system-wide inputs. Real-time network traffic analysis using the fuzzy inference system (FIS) is also beneficial. FIS is effective at identifying network intrusions because it has excellent classification accuracy and rapidly trains datasets (Pannu 2015, 79–80).

Concept of machine learning

Machine learning has changed significantly throughout the years. It is one of the foundational elements of information technology, and as a result, it has grown to be a very important—though sometimes overlooked—aspect of daily life. It is without a doubt the technology that consumers are most interested in today. Applications include autonomous vehicles and the prediction of fatal diseases like Amyotrophic Lateral Sclerosis (ALS). Given the growing amount of data available, there is strong evidence that smart data analysis will become more common and important for technological advancement.

Since the technical revolution, an immeasurable amount of data has been generated. Studies have shown that around

- Quintillion bytes of data are created daily (Sivarajah, Muhammad, Zahir, and Vishanth 2017, 265). These data come from different sources, such as the cloud, the Internet of Things (IoT), social media platforms, and so on. The increased usage of the internet has skyrocketed, which has in turn led to a high amount of data. Since there is too much data, there is a need to find a suitable method that can analyse, process, interpret, and make sense of this data, as stated on Nick's website on October 27, 2022. The most suitable way is called machine learning. Businesses like Amazon, Netflix, Facebook, Instagram, and so on use machine learning to increase profit and reduce risks. This is also one of the benefits of machine learning; it is applicable for predicting risks, maximising profits, and identifying opportunities (Corbett-Davies & Goel, 2018).

It is essential to first fully understand what machine learning entails. Artur Samuel originally used the phrase "machine learning" in 1959. In its simplest form, machine learning is a set of methods for teaching computers to perform tasks that humans traditionally perform better than machines. This is accomplished by educating computers to learn like people. In the context of a machine, machine learning is a subfield of artificial intelligence that focuses on developing systems that can learn from experience and make predictions using data. This method enables machines to think, which allows them to solve problems. Using training data or prior knowledge, a computer programme can be used to optimise a model's parameters. Models are specified up to a point, and learning is the use of models to enhance efficiency. It involves using sample data or foreknowledge, programming computers to maximise a performance criterion, and learning. The model may be descriptive to help make inferences from the data or predictive to look ahead. Typical examples of machine learning utilisation in our immediate environment are image recognition, speech recognition, medical diagnosis, and so on.

The practical scenario of machine learning in business

Various organisations, companies, and businesses make use of machine learning. A few common ones are:

- Netflix: Netflix is a video streaming company. It uses machine learning to suggest movies to its users depending on their preferences. Additionally, it uses machine learning to suggest movies to viewers based on the genres of previously seen movies.
- Google Mail: Google Mail, also known as Gmail, uses machine learning and natural language processing to analyse mail. This helps to differentiate between real mail and spam mail.

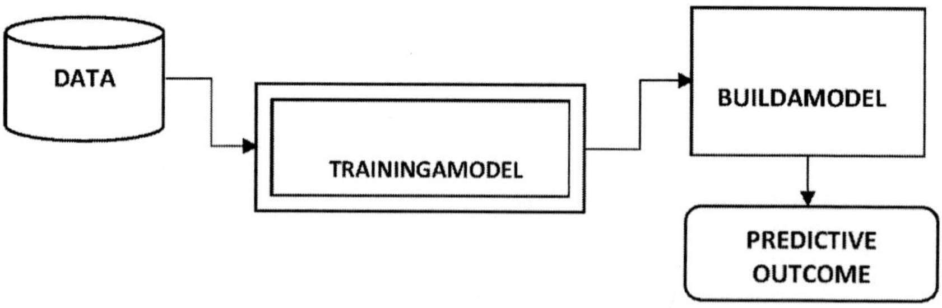

Figure 1: Basic Concept of Machine Learning

Need for machine learning

There is no doubt that, in this age and time, machine learning is essential. Below are some of the reasons why machine learning is essential in society.

1. More data is being generated.
2. Enhance decision-making
3. Uncover trends and patterns in the data.
4. Solves complex problems

Characteristics of Machine Learning

The following can be said to be some of the attributes of machine learning:

1. Automated data visualisation capabilities
2. Automation done right
3. the capacity to combine efficiency with IoT to increase productivity
4. precision in data analysis
5. enterprise intelligence

Classifications of Machine Learning

Machine learning is of primarily three (3) types: supervised, unsupervised, and reinforcement learning. These are conceptualised and explained further below.

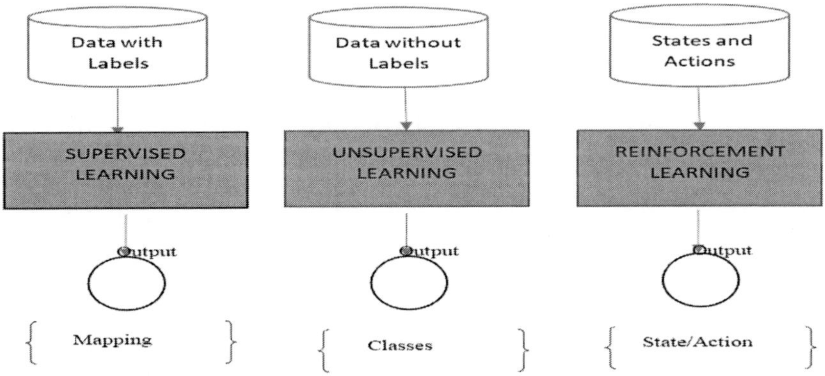

Figure 2: Classifications of Machine Learning Algorithm

Supervised Learning

Supervised learning, as the name suggests, involves working under supervision. Simply said, it is a form of machinelearning that resembles learning from a teacher. To discover the knowledge about the relationship between input and output in a system, a collection of paired input-output training samples is required. Since the output is regarded as the label of the input data or the supervision, an input-output training sample is sometimes referred to as labelled training data or supervised data. Another way to define supervised learning is as a model that can make predictions using data that has been labelled. A labelled dataset in machine learning is raw materials, including pictures, texts, videos, etc, that can be identified and give the context that permits the addition of one or more labels that are in formative and instructive so that the model of machine learning can learn something.

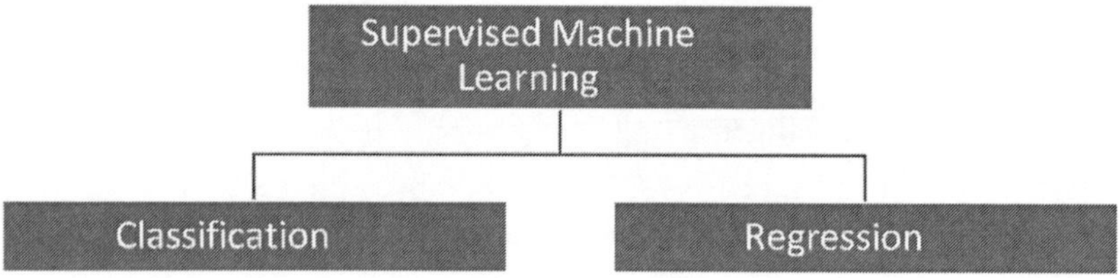

Figure 3: Classification of Supervised Machine Learning Algorithm

Classification and regression are further divisions of supervised learning. When a categorical output variable—one with two (2) or more classes—is utilised, such as yes/no, true/false, or male/female, it is classified. Random forests, decision trees, logistic regression, and support vector machines are supervised machine learning methods. Using a categorical target or categorical class variable to predict related information is known as classification. Any kind of statistical data can be gained from using this tactic. Numerous applications, such as image classification, predictive modelling, data mining, etc., make use of these techniques. Building a straightforward and unambiguous model of how class labels are distributed among predictor features is the basic goal of supervised learning. Using the classifiers, the value of the unknowable class label is then compared to the class labels of the testing cases, where the values of the predictive features are known. Regression is a supervised learning technique that makes it possible to find correlations between variables and use those correlations to predict continuous values. It is a regression when a real or continuous value is the output variable, that is, the relationship between two or more variables where a change in one variable is associated with a change in other variables (salary based on work experience). Techniques in supervised machine learning are linear regression, regression trees, nonlinear regression, Bayesian linear regression, and polynomial regression (Salim 2021, 5-9).

Procedure for Supervised Learning: Labelled Data

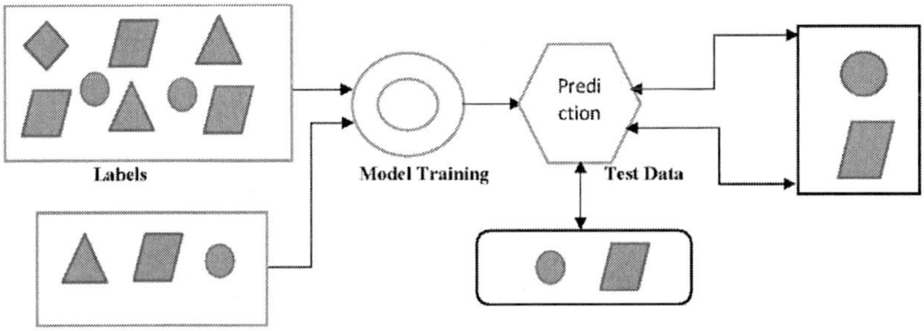

Figure 4: Concept of Supervised Machine Learning

Algorithm Pseudocode for Supervised Learning

1. Identify the type of training dataset.
2. Using labels, collect the training.
3. From the training dataset, produce tests, validation, and training.
4. Identify the training dataset's input features, which need to be sufficiently specific to enable reliable output.
5. Select the approach that works best for the model, such as a support vector machine or a decision
6. Use the practise data to test the It is occasionally necessary to use validation sets, a subset of training datasets, as control parameters.
7. Assess the model's accuracy using the test. The model is considered accurate if it accurately predicts the result.

Merits of Supervised Learning

1. The model can anticipate the outcome based on prior experiences with the aid of supervised machine learning algorithm.
2. The object classes can be known via supervised learning algorithm.

3. Spam filtering and fraud detection are two examples of the real-world problems we tackle using supervised learning.

Demerits of Supervised Learning

1. Models of supervised learning are inadequate to handle challenging real world problems.
2. Supervised learning cannot accurately predict the outcome if the test data and training dataset are incomplete.
3. The training involved lengthy computations.
4. To do supervised learning, one must have sufficient information on the object.

Unsupervised machine learning

Unsupervised machine learning techniques allow the algorithm to act on the data without supervision and are used to train the algorithm using data that is either unlabelled or unclassified. Unlabelled data is a term used to describe information that has not been labelled, indicating traits, qualities, or classification. Unsupervised learning begins with training data that is based just on the feature vector and has no labels. The prediction model developed by the machine learning model determines the potential or cluster ID, or a better example, of unlabelled data, and this is verified using a specific subset of data and its feature vector. Unsupervised learning cannot be utilised to quickly resolve a regression or classification problem, in contrast to supervised learning, where we know the input data but have no corresponding output data. Unsupervised learning tries to reveal a dataset's underlying structure, categorise the data into relevant groups, and present the dataset succinctly.

Reasons for Unsupervised Machine Learning

Below are some of the major advantages of unsupervised learning over supervised learning.

1. Unsupervised learning facilitates the discovery of valuable insights from data.
2. It more closely resembles how humans learn to think through their own experiences, which brings it closer to actual artificial intelligence.

3. Unsupervised learning is more significant because it operates on unlabelled and uncategorized data.
4. Unsupervised learning is necessary to handle situations where the input and output are not always the same in the real world.

Unsupervised machine learning techniques can be divided into two major types:

1. Clustering
2. Association

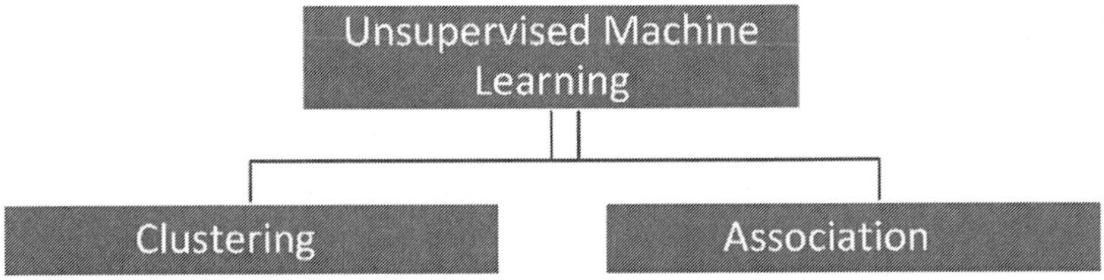

Figure 5: Types of Unsupervised Machine Learning Algorithms

Clustering

Clustering is a method of grouping comparable groups of uncategorized data. This procedure makes sure that linked data points are found and grouped. Clustering, the biggest unsupervised learning issue, entails finding a structure in a collection of unlabelled data. It is the process of grouping objects into sets whose members share characteristics. As a result, a cluster is a group of objects that are distinct from those in other clusters but similar to one another (Salim 2021, 5–9). Algorithms for clustering data are essential for processing it and identifying groups (natural clusters).

Types of clustering in machine learning are:

1. K-means Clustering
2. Hierarchal clustering

Association

An association rule, a type of unsupervised learning technique, is used to discover associations between the variables in a large database. It indicates the group of items in the collection that commonly occur together. The association rule is a type of unsupervised learning in which the algorithm tries to learn without a tutor because the data are not labelled. The discovery of fascinating linkages hidden in massive datasets is often accomplished using association rules, which are descriptive rather than predictive. Typically, the link is denoted by rules or often occurring item sets (Salim 2021, 5–9).

Pseudocode for Unsupervised Machine Learning (case of k-means clustering)

1. Decide the number of clusters represented by the value K.
2. Select m random data points from the dataset as the centre of clusters (referred to as the centroid), where m = k.
3. Allot the data points to the nearest cluster centroid.
4. After assigning each data point to a particular cluster, recalculate the centroid.
5. Repeat steps 3 and 4 till the maximum number of iterations is reached (the number of iterations is set by the user).

Advantages of Unsupervised Machine Learning

1. Because it doesn't need labelled input data, unsupervised learning is utilised for more complex problems than supervised learning.
2. Unsupervised learning is advised because it is simpler to get unlabelled data than labelled data.
3. Because it may aid in the interpretation of raw data, unsupervised learning is the finest tool for data.
4. Additionally, it can establish how comparable the data is. To do this, probabilistic approaches can be applied.

Disadvantages of Unsupervised Machine Learning

Unsupervised learning is intrinsically harder than supervised learning because there isn't a comparable outcome. The results of the unsupervised learning approach may be less accurate because the input data is not labelled and the algorithms are unable to forecast the precise result in advance. As the number of features increases, the complexity increases.

Reinforcement machine learning techniques

Simply explained, reinforcement machine learning is a category of machine learning in which an agent learns how to behave in a certain environment by doing it and then observing the results. Reinforcement learning allows an AI agent (a software component) to automatically explore its surroundings by striking and trailing, acting, learning from experiences, and improving performance. The basis for reinforcement learning is feedback. Since the agent is rewarded for every positive behaviour and penalised for every negative action, the objective of a reinforcement learning algorithm is to increase rewards. Reinforcement learning differs from supervised learning in that it is solely based on the experiences of the agents.

Types of reinforcement machine learning techniques

1. Positive reinforcement machine learning technique
2. Negative reinforcement machine learning technique

Positive reinforcement machine learning technique: By adding something, positive reinforcement learning refers to raising the likelihood that the desired behaviour will occur once more. It strengthens the agent's behaviour and has a favourable effect on it.

Negative reinforcement machine learning technique: A negative situation that may have been stopped or avoided leads to reinforced behaviour. Negative reinforcement is what is meant by this. It can be utilised to define the minimum performance requirement. The disadvantage of this strategy is that it only offers sufficient incentives to motivate the barest amount of activity.

The usefulness of the reinforcement machine learning technique:

1. It aides in identifying situations that require intervention.
2. It identifies the action that produces the greatest reward.
3. The learning agent also has a reward function to calculate the reward value based on each action.
4. It also enables it to determine the most effective method for generating substantial returns.

Reinforcement learning is computationally expensive and time-consuming if you have enough data to solve the problem with a supervised learning method, especially if there is a huge action space.

Limitations of reinforcement Machine learning techniques

Listed below are some of the limitations that affect machine learning with reinforcement:

1. Learning rate may be affected by partially observable environments in the actual world.
2. The findings may be compromised by an abundance of states resulting from excessive
3. Realistic environments can evolve over time.

Connectivity between the ML and AI algorithms

In discussing the connectivity between the ML and AI algorithms, a summary of the connectivity between ML and AI is necessary. This connectivity spans several areas. ML is a subspace of AI and is generally employed to make smart intelligence out of machines to enable them to learn intelligence automatically based on their intelligent learning experience. For data processing systems, ML has emerged as a cutting-edge technology that provides autonomous operations and decision assistance. In the area of autonomous driving, the connectivity of ML and AI is manifested, as both AI and ML form the basis of the technology that enables computer-driven cars to navigate from point A to point B and vice versa. The overview of this connectivity reflects that ML helps build AI-driven applications.

All ML algorithms, whether supervised or unsupervised, require data in their implementations. They all solve problems without human intervention. ML algorithms are used to automatically understand and realise the day-to-day problems that people are facing. Some algorithms can be used for the same purpose, while others are for different purposes. In most cases, no algorithm is considered perfect; some algorithms are better at solving problems using unlabelled data while others do well with labelled data. Reinforcement learning also encompasses the direct acclimatisation and exploratory activities meted out on functional data or situations present within the space as an agent of an AI system.

Access Control

Any process by which a system provides or revokes the right to access some activity is referred to in this chapter as access control. A user frequently has to log in to a system using authentication. Authentication is the process by which a person or computer verifies their identity with a server or control system. In other words, access control procedures are the methods and mechanisms used by information owners to grant permission for users to access data, information, and systems. Typically, a server will deploy a username and password as part of the authentication process. Card authentication, retina scans, speech recognition, fingerprint authentication, and other methods are also possible. The authentication procedure consists largely of three parts, namely:

1. Identification of the user (who is it?).
2. Provision of some kind of authentication information that is secret and unforgettable.
3. Transmission of the authentication information to the system through a secure.

The authentication scheme largely hinges on the security of sensitive and relevant data, with procedural components of the features making up the security system. Access control is one of the most fundamental security measures, though. The access control process is divided into five main tiers, which are as follows:

Authorization: This involves granting access or permission to the user of the common resources.

Authentication: This involves the user identifying a piece of information using pre-defined schemes, for instance, username and password.

Accessing: This involves enjoying direct access or interfacing with corporate resources (data and information).

Management: Effective utilisation of the available resource.

Auditing: Evaluation of the transaction motives and activities embedded in access control procedures.

In the end, access control is a notion that controls who or what can access resources in a computing environment to prevent all types of incursions. It is an essential component of security that reduces risk to the success of organisational or company resources. Notably, there are two different methods of access control.

They are:

1. Physical access control
2. Logical access control

Using cards and control panels, physical access control restricts access to campuses, buildings, rooms, and physical IT assets where resources in the form of data or information are housed in computing facilities. Physical access control explicitly defines the stringent physical space that supports or reinforces access limits. Logical access control restricts access to data, system files, and computer networks. By analysing the necessary log credentials, the logical access control system performs identification (authentication) and authorization of people and entities.

Data

There is no denying that data is a precious resource, and without question, in this contemporary electronic age, data is a very important commodity. The digital world is incredibly rich in data, including data on social media, business, cybersecurity, and other topics. To carefully and efficiently analyse these data and create a connection between the data and real-world applications, it is necessary to employ artificial intelligence expertise and ideas, namely machine learning. Several machine-learning approaches can examine data and forecast events in the real world. Different types of data are available for analysis. Data can typically be classified as structured, semi-structured, or unstructured.

Structured data: Its structure is clearly described, it complies with a set of guidelines for data organisation and retrieval, and it is used by an individual, team, or computer programme. In well-defined systems like relational databases, the structured data is frequently stored and presented in tabular format. Names, dates, places of residence, credit card information, stock levels, and geolocation are examples of structured data (Rusu, Ionela, Oana, and Giorgianet al.2013, 2).

Unstructured data: As opposed to that, unstructured data, which primarily consists of text and multimedia content, has a predefined framework or organisation and is therefore far more challenging to capture, handle, and analyse. Sensor data, emails, blogs, wikis, word processing papers, PDF files, audio files, videos, photos, presentations, and web pages are a few examples.

Semi-structured data: Semi-structured data is not stored in a relational database like the aforementioned structured data, but it does have some organisational qualities that make it easier to study. Things like HTML, XML, JSON files, NoSQL databases, and more include semi-structured data (Rusu, Ionela, Oana, and Giorgian et al.2013, 2).

Structured data are typically much easier to analyse than unstructured data; however, that is by no means to put aside information gotten from unstructured data; details gotten from unstructured data can give a much more comprehensive insight into the analysis. There are numerous techniques to get unstructured data, such as Internet scraping, data extraction from human languages using natural language processing, the use of NoSQL databases, etc. The merits of unstructured data analysis are clear, given that the vast bulk of the information we have access to is unstructured. It can supply an abundance of insightful information and suggestions on how to enhance the effectiveness of a business or a particular service. Making the data "understandable" for computers is the initial action if we want a machine to process it. The distance between people's comprehension and computer processing needs to be bridged. It indicates that the relevant data is typically processed manually by a human operator before being converted to a format that can be processed by a computer.

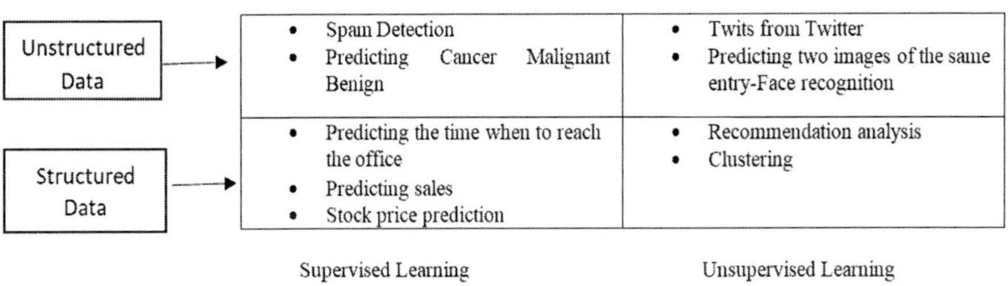

Figure 6: Comparison between structured and unstructured data

Data labelling

The act of attaching labels or tags to unprocessed data, such as images, movies, content, and audio, is referred to as data labelling. These tags serve as an illustration of the entity class to which the data belongs and aid machine learning algorithms in learning to recognise that particular object class when it appears in data without a tag. Data labelling approaches are:

1. In-house data labelling
2. Crowding sourcing

3. Outsourcing
4. Automatic data labelling

Data assets are being used by organisations and businesses more and more to boost productivity and competitiveness. Utilising machine learning tools and technology is a component of this strategy. However, compared to general workloads, machine learning workloads have quite distinct computing and data storage requirements.

Machine learning applications need a lot of data to build, train, and maintain the models. The most important factors to take into account for the storage of these applications are highly efficient and long-term data storage (Arulmozhi Pavithra and Sivasankari 2017, 782–783). Organisations and businesses can leverage existing data sources to create and train models that produce insights to enhance corporate operations, more precisely target customers, or generate better goods. When developing storage, it is important to observe the requirements, which are:

1. Scalability
2. Accessibility
3. Latency
4. Throughput
5. Parallel access

These are the kinds of data that our stores need to be well protected. Data security is an essential factor in data storage. Different security measures can be put in place for data security; access control is onc of the key security elements that can be used. Access control is what defines a segment of data that internal parties can see, and the larger an organisation gets, the more important access control gets. Access control is useful not only to protect sensitive data but also to ensure that different sectors in the organisation only get data that is relevant to them. This helps to reduce noise, and being able to isolate data through security is very essential.

Access control is a method for confirming that users are who they claim to be and have permission to access business data. Access control is essentially the purposeful limitation of data access (Aissam, Anas, Hamza, and Siham 2020, 333–336). Two essential components are authentication and authorization. Authentication is a process for verifying that someone is who they claim to be. Although authentication alone is insufficient to safeguard data, it can be used in conjunction with permission to decide whether a user should be permitted to access the data or complete the transaction they are attempting (Athira and Lekshmy 2015, 99).

Integrated Individualised Organisation Resource

Individual organisations require a maximum volume of data as a string indicator or factor of production to accomplish most of their corporate objectives. It is therefore incumbent on the organisation to consider the cost of storage, management, and analysis of most of its data and other data processing tasks. The fusion of various enterprises with similar data needs (as conceptualised in Figure 7) will help them to operate optimally at a reasonable cost of data storage, management, and other data manipulation activities. The database, which largely comprises logically and functionally aggregated datasets that lean on the fact that it is a sharable resource and provides a remarkable fulcrum for solving most organisational challenges and tasks in the emerging knowledge society, is a nutritional recipe needed for the flourishing of most modern enterprises. On the strength of the paramount importance of datasets, which are technically stored in a good space, it can promote the well-being of such enterprises and also allow for their upward growth and attainment of lofty achievements. The scenario where these individual enterprises now forge massive alliances provides the platform and necessary antidote to explore the immeasurably high benefits accruable from a robust integrated individualised enterprise 'resource,' providing a new order of driving modern business activities that is neat and economical for significant productivity and profit. In terms of intrinsic qualities inherent in the various individualised enterprises, there are possibilities for common convergence of some of the data resources of these organisations and meeting their corporate objectives and attainment with relative ease. It is quite fascinating to perceive the integration of similar enterprises' data resources for the progress of the global economy as a new vista to the liberation of the global economy. The network security issue, data analysis, data manipulation, data storage, and usage are fundamental grounds that require efficient and effective management for the ultimate productivity of these organisations. The integrated resource can interface or converge to upstage some global needs for data resources available within the foreseeable future.

Artificial Intelligence in Databases

There are quality databases for AI and machine learning. The game has changed recently as a result of AI's role in databases. By gradually integrating AI, the problem of database technology's constrained modelling capabilities has been overcome, and it is believed that this integration will have a big impact on the future of computing. The necessity for access to vast volumes of shared data for data processing, efficient data management, and intelligent data processing are the primary driving forces for the integration of these two technologies. The information technology (IT) sector is creating autonomous hardware and the first autonomous databases, intending to have artificial intelligence (AI), machine learning, and sophisticated programme algorithms execute the majority of maintenance tasks automatically (Shvertner 2020, 305–307). As seen in numerous instances when AI approaches provided semantic aid to database systems, AI systems can efficiently manage to process a vast amount of data and information (Catania and Zarri 2000, 34). Metadata representation, integrating heterogeneous sources, online application design, access control, and authentication are a few significant applications of AI database integration.

In some cases, the large volume of data in a database can be realised as a result of a repository housing several databases from different organisations. This is what is referred to as an integrated enterprise database (IED). It is expedient, considering the commanding influence of AI and ML-driven technologies today, to explore the avalanche of benefits, which can be derived from IED with a common or almost similar resource for driving their various organisations' activities. It is quite amazing technologically to build an AI-inspired database to speed up a machine learning model for making quality predictions for the accomplishment of organisational objectives at a relatively low cost, in less time (milliseconds), and with seamless access to an integrated enterprise resource saved in the cloud. However, it is anticipated that such a database will assist in more effectively addressing volume, velocity, and complicated data governance and management concerns that are related to training machine learning to reduce training time and improve access and access control. To provide services aimed at maximising the use of computer and database resources, AI databases incorporate AI technology. In a matter of milliseconds, AI databases can consume, examine, analyse, and visualise a data source on a computer.

Cloud data storage

Cloud data storage is a service that permits users to store data on a remote storage system that is overseen by a different party and made accessible through web services. Technically, it means stashing data on hardware located physically far away that can be accessed via the internet; in this case, customers upload files to a server run by a third party. A typical example is Dropbox. Access control is a method for confirming that users are who they claim to be and have permission to access business data. Access control is essentially the purposeful limitation of data access (Aissam, Anas, Hamza, and Siham 2020, 333–336). Two essential components are authentication and authorization. Authentication is a process for verifying that someone is who they claim to be. It is only a made-up word for a constructed virtual space in the cloud. Users' information may be stored on any machine in a data centre that has many locations. The actual storage location of the data may periodically change since the cloud dynamically maintains accessible storage sites close to the data centres. The user can handle the storage space as if they were on a personal computer, even though the data location is virtual and the user sees a static location for their data. A typical cloud data storage system is depicted in the figure below.

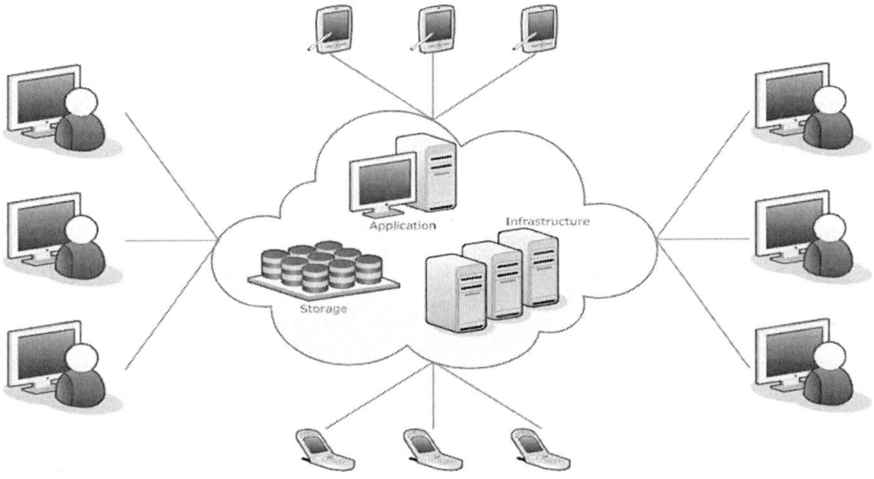

Figure 8: A Cloud Data Storage

Data storage in the cloud is distinguished by the following characteristics:

1. It consists of numerous remote resources yet still functions as one, also known as federated storage clouds.
2. Data redundancy and dissemination make it very fault-tolerant.
3. Versioned copies increase reliability, however when it comes to data replicas, they typically result in information mirroring.

The following can be said to be some key issues relating to cloud data storage:

1. Deployment of cloud storage
2. Data de-duplication
3. Security
4. Data migration and load balancing
5. Data organisation

The figure below depicts a flow of algorithms for how users can secure and access data on data storage (cloud/authentication scheme).

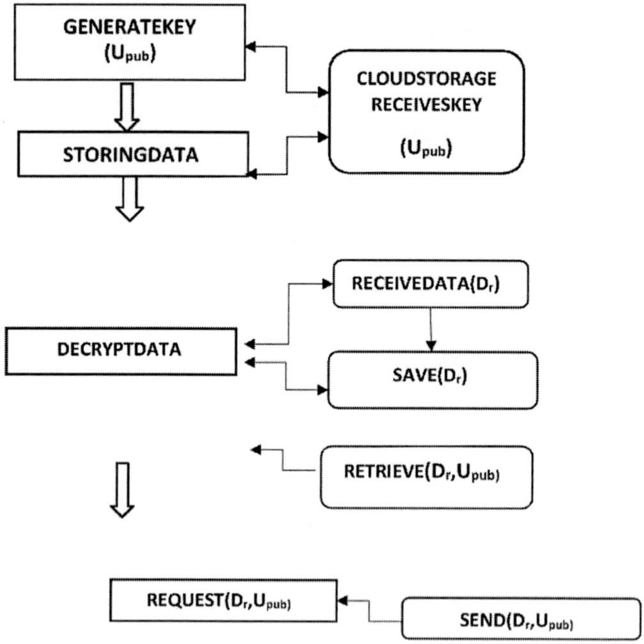

Figure 9: Algorithm for Cloud Data Storage

Where **DR represents data,** Upub represents the generated key for the user.

Machine learning is without a doubt playing a significant role in database management. Even some databases are embedded with machine learning techniques. The Machine Learning Database, sometimes known as MLDB, is an open-source system made to handle machine learning tasks involving large amounts of data. It can be utilised to collect and store data for machine learning model training or to deliver real-time prediction endpoints. Large-scale businesses, large organisations, or companies can greatly benefit from machine learning in databases. The benefits of machine learning techniques in the database are that it brings about simplicity and a reduction in time; it also implies that you may reduce the number of steps necessary to implement machine learning that is more effective, quicker, and simpler with the correct database and machine learning supplier (Sai 2020, 25137).

Complexity of the overall process

Time complexity is characterised in terms of the number of times an algorithm must be executed, depending on the size of the input. Because programming language, systems, and pre-processing power are also taken into account, time complexity is not a measure of how long it takes to execute a certain method. A form of computational complexity called time complexity specifies how long it takes to run an algorithm. The length of time required for each statement to finish execution is known as the algorithm's time complexity. It therefore heavily depends on the volume of the processed data. Additionally, it helps define an algorithm's effectiveness and assess its performance. The link between time complexity and input size is significant. The runtime, or the length of time it takes the algorithm to run, increases as the size of the input does.

Considering the entire process in the framework in Figure 7, the time complexity will be described in two parts. The first is the time complexity for all the individual organisation resources to be integrated into the AI-driven database. For the case scenario of a hundred to thousands of organisations, the merge sort algorithm is implemented. The time complexity can be expressed as $T(n) = 2T(n/2) + O(n)$, with the solution given as $O(n*\log n)$, where n denotes the number or size of resources being integrated. For the case scenario of only a few tens to hundreds of organisations, the heap sort is implemented with $O(n\log n)$ as the time complexity for the best case, average case, and worst case. It should be noted that heap sort is to be considered for smaller sets, while merge sort is to be considered for larger sets. The second part is the process of cloud storage with an authentication scheme utilising the RSA (Rivest-Shamir-Adleman) algorithm. The time complexity in the general case is O log 2. But in common practical use, the time complexity is $O(n^3)$ when a private key is in use and $O(n^2)$ for public key use. The use in both cases covers signature generation, verification, encryption, and decryption.

Artificial Intelligence in Access Control, Authorization, and Authentication

Sensitive data held by an organisation must have access control over any database to define the level of access granted to different authorised users and to prohibit access to unauthorised users. Two key elements of access control are authentication and authorization, which confirm user identification for any database, and the level of access. However, a dynamic access control system with an AI implementation is crucial for protecting the cloud-based database. This will ensure and create a strong structure that will be challenging for intruders and interlopers to breach. For managing access to Google Cloud resources, for instance, Google utilises Identity and Access Management (IAM), a solution powered by artificial intelligence (Google Cloud, 2021).

Figure 10 below shows a code snippet sample from a JSON file in Python that handles the authentication of users to cloud storage.

```
{
  “cloud_source”:“google”,
  “Server”:http://googelcloud.org/developers/logger/access-
  conrtol,“AUTH_TOKEN”:“3gh54tr”
}
```

Figure 10: Code snippet for user authentication

The technologies that are provided by AI and machine learning stand the chance of ensuring that access to databases is better fortified to put a proper check on the identity of the user and to also determine the level of access to grant users, since users in some organizations may have more than one role.

Conclusion

The available information on data generation as of 2015 revealed that enterprises recorded a total of 18.2 zettabytes of data. It has also been projected that the total volume of data generated by the global enterprise will attain a total volume of 175.8 zettabytes (which is equivalent to one trillion gigabytes) in the year 2025. From the volume of data generated currently, only about one-third is available in a good state for comprehensive analysis to be conducted, while others are lying untapped. It should be noted that data volume is increasing at a relatively faster pace than enterprises can keep up with, and enterprises are most often confronted with how to effectively store, manage, and analyse data. The utilisation of machine learning, nonetheless, will continue to enhance, improve, and significantly modify the various aspects of design issues in the database as well as query optimisation. Here, authentication and access control issues will attract considerable attention to enterprises' resources, and this will greatly continue to support the growth of data volume for the overall well-being of enterprises' operational efficiency This work is expected to accommodate, in the near future, the implementation of the proposed conceptual framework, where the major challenges confronting data management using AI technologies and machine learning are considered. These problems include data indexing, the expense of joining tables, and the computational cost of copying data into another repository for analysis. Research efforts are capable of contending with manipulative efforts geared towards exploiting datasets for positive enhancement and enterprise optimally conceived performance desirable towards global economic breakthroughs. Available machine learning techniques, which are AI-strengthening, should be continually visited and polished for an overall improvement in global enterprise productivity, the hallmark of doing business in any environment worldwide.

Authentication checks and access control on integrated enterprise databases (IED) form a veritable point of serious concern at the technical level. These two key attributes are vastly needed to keep the IED alive and healthy. The preservation of these unified resources (sharable) from interlopers and other network scavengers will no doubt prolong the existence and long-term benefits of the technology for institutional individuals, collective databases, and integrated enterprise repositories. Thus, the linked databases form shareable resources that are accessible to arrays of transactional activities based on specific algorithmic steps.

References

1. Akhila Raj and Angel Viji 2018. "A Survey on Data-Centric Access Control with Role-Based Proxy Re-encryption in a Cloud Environment", *International Journal of Scientific Engineering and Research,* 7, no. 3: 31–35.
2. Aissam Outchakoucht, Anas Abou El Kalam, Hamza Es-Samaali, and Siham Benhadou 2020. "Machine Learning-Based Access Control Framework for the Internet of Things." *International Journal of Advanced Computer Science and Applications,* 11, No. 2, 331-340.
3. Anwar, Amaan, and Syed Imtiyaz Hassan 2017. "Applying artificial intelligence techniques to prevent cyber assaults." *International Journal of Computational Intelligence Research,* No. 5: 883–889.
4. Arulmozhi, Bavita, Pavithra, Javind, and Sivasankari Ashirat (2017) "Cloud Computing Environment Using Secured Access Control Technique." *International Research Journal of Engineering and Technology*, 4, no. 12: 780–788
5. Athira Ramos and Lekshmy Kumar 2015. "Secure Data Storage in the Cloud Using Centralised Access Control with Anonymous Authentication. *International Journal of Engineering and Advanced Technology* (IJEAT), 5, no. 1: 98–101.
6. Avram, Maricela-Georgiana. 2014. "Advantages and challenges of adopting cloud computing from an enterprise perspective." *Procedia Technology,* No. 12: 529–534.
7. Bekhit, Ahmed Fathi, and Ahmed Fathi Bekhit. 2022. "Introduction to Computer Vision." *Computer Vision and Augmented Reality in iOS: OpenCV and ARKit Applications*, 1–20
8. Catania, Barbara, and Gian Piero Zarri 2001. *Intelligent database systems* Boston: Addison-Wesley.
9. Corbett-Davies, Sam, and Sharad Goel 2018. "The measure and mismeasure of fairness: A critical review of fair machine learning." *arXiv preprint arXiv:1808.00023*.
10. Cuzzolin, Fabio, Alice Morelli, Bogdan Cirstea, and Barbara J. Sahakian 2020. "Knowing me, knowing you: theory of mind in AI." *Psychological Medicine*, No. 7: 1057–1061
11. Dridi, Salim. 2021. "Supervised and Unsupervised Machine Learning: A Systematic Literature Review." *A Preprint,*
12. 1-22.

13. Gill, Karamjit S. 2016. "Artificial superintelligence: beyond rhetoric." *AI & Society,* No. 31: 137–143.
14. Google, Cloud, 2021. "Identity and Access Management (IAM)". Accessed October 5, 2022. https://cloud.google.com/iam.
15. Goralski, Margaret A., and Tay Keong Tan 2020. "Artificial intelligence and sustainable development." *The International Journal of Management Education,* 18, No. 1: 78–85 https://doi.org/10.1016/j.ijme.2019.100330.
16. Gurjeet, Singh, and Mohita Garg 2018. "Data Storage in Cloud Computing: A Review". *International Journal of Computer Science and Technology*, 17, no. 2: 7206-7214
17. Gusmeroli, Sergio, Salvatore Piccione, and Domenico Rotondi 2013. "A capability-based security approach to managing access control in the Internet of Things." *Mathematical and Computer Modelling,* 58, No. 5–6: 1189–1205.
18. Hecht, Jeff. 2018. "Lidar for self-driving cars." *Optics and Photonics News,* 29, No. 1: 26–33.
19. Jamieson, Rodger, and Meliha Handzic 2004. "A framework for security, control, and assurance of knowledge management systems" *Handbook on Knowledge Management 1: Knowledge Matters,* No. 8: 477–505.
20. Joshi, Shubham, Shalini Stalin, Prashant Kumar Shukla, Piyush Kumar Shukla, Ruby Bhatt, Rajan Singh Bhadoria, and Basant Tiwari 2021. "Unified authentication and access control for future mobile communication-based lightweight IoT systems using blockchain" *Wireless Communications and Mobile Computing,* No. 8: 1–12.
21. Kelley, Troy, and Lyle Long (2010) "Deep Blue cannot play checkers: the need for generalised intelligence for mobile robots." *Journal of Robotics*, No. 10: 58–66.
22. Kuusi, Osmo, and Sirkka Heinonen. 2022. "Scenarios from Artificial Narrow Intelligence to Artificial General Intelligence—Reviewing the Results of the International Work/Technology 2050 Study." *World Futures Review,* 14, No. 1: 65–79.
23. Laskar, Sahinur Rahman, Abdullah Faiz Ur Rahman Khilji, Partha Pakray, Rabiah Abdul Kadir, Maya Silvi Lydia, and Sivaji Bandyopadhyay 2022. "Heal Favor: A Machine Translation-Enabled Healthcare Chat-Based Application." In *Natural Language Processing in Healthcare*, 121–137. Ohio: CRC Press.

24. Lieto, Antonio, Mehul Bhatt, Alessandro Oltramari, and David Vernon. 2018. "The role of cognitive architectures in general artificial intelligence." *Cognitive Systems Research,* No. 48: 1–3.
25. Lu, Huimin, Yujie Li, Min Chen, Hyoungseop Kim, and Seiichi Serikawa 2018. "Brain intelligence: go beyond artificial intelligence." *Mobile Networks and Applications,* No. 23: 368–375
26. Maraju, Kinsha. 2018. "Applying AI in Application Security." *Journal of Information Systems Audit and Control Association*, No. 1: 1–7.
27. Messika, Ben. 2018. "Cryptocurrency and Blockchain: The Future.". Accessed August 29, 2022. https://medium.com/theblock1/cryptocurrency-andblockchain-the-future-9cd689e977ca
28. Memon, Muniba, Syed Shahbaz Hussain, Umair Ahmed Bajwa, and AsadIkhlas. 2018. "Blockchain beyond Bitcoin: Blockchain technology challenges and real-world applications." In the *2018 International Conference on Computing, Electronics, and Communications Engineering (iCCECE)*, 29–34
29. Mohammed, Ishaq Azhar. 2021. "The interaction between artificial intelligence and identity and access management: an empirical study." *International Journal of Creative Research Thoughts, ISSN* 2320, No. 2882: 668–671.
30. Mohammed, Ishaq Azhar. 2012. "Analysis of Identity and Access Management Alternatives for a Multinational Information-Sharing Environment." *International Journal of Advanced and Innovative Research*, No. 8: 1–7.
31. Nick Health, 2020. "What is machine learning? Everything you need to know." Accessed October 27, 2022. https://www.zdnet.com/article/what-is-machine-learning-everything-you-need-to-know/
32. Pannu, Avneet. 2015. "Artificial intelligence and its application in different areas." *International Journal of Engineering and Innovative Technology*, No. 10: 79–84.
33. Picard, Rosalind W., and Jonathan Klein. 2002. "Computers that recognise and respond to user emotion: theoretical and practical implications." *Interacting with computers,* 14, No. 2: 141–169
34. Priya, B., J. M. Nandhini, and T. Gnanasekaran 2021. "An Analysis of the Applications of Natural Language Processing in Various Sectors." In *Smart Intelligent Computing and Communication Technology*, 598–602, Amsterdam: IOS Press.

35. Pueyo, Salvador. 2018. "Growth, degrowth, and the challenge of artificial superintelligence." *Journal of Cleaner Production,* No. 197: 1731–1736
36. Reddy, Ravi Vijaya Kumar, and U. Ravi Babu 2018. "A review on classification techniques in machine learning." *International Journal of Advance Research in Science and Engineering,* 7, no. 3: 40–47.
37. Rusu, Octavian, Ionela Halcu, Oana Grigoriu, Giorgian Neculoiu, Virginia Sandulescu, Mariana Marinescu, and Viorel Marinescu 2013. "Converting unstructured and semi-structured data into knowledge" In 2013, the *11th RoEduNet International Conference,* 1-4
38. Saleh, Ziyad. 2019. "Artificial Intelligence Definition, Ethics, and Standards". *Journal of Artificial Intelligence*, No. 14: 870–892.
39. Sai Tanishq, 2020. "Machine Learning for Database Management Systems." *International Journal of Engineering and Computer Science, 9, no. 88: 25132–25147.*
40. Schuld, Maria, and Francesco Petruccione (2018) *Supervised learning with quantum computers*. Berlin: Springer.
41. Seethamraju, Ravi. 2015. "Adoption of software as a service (SaaS) enterprise resource planning (ERP) systems in small and medium-sized enterprises (SMEs)." *Information Systems Frontiers,* No. 17: 475–492.
42. Shvertner, Krassimira. 2020. "Artificial Intelligence Approaches in Database Management Systems," Yearbook of the Faculty of Economics and Business Administration, Sofia University, Faculty of Economics and Business Administration, Sofia University St Kliment Ohridski, Bulgaria, 18, No. 1: 303–326.
43. Singh, Gyanendra, and Sagar, Dheeraj 2013. "An Overview of Artificial Intelligence." *SBIT Journal of Sciences and Technology,* 2 No. 1: 56–69
44. Sivarajah, Uthayasankar, Muhammad Mustafa Kamal, Zahir Irani, and Vishanth Weerakkody 2017. "Critical analysis of big data challenges and analytical methods." *Journal of Business Research,* 7: 263-286.
45. Sridharan, Singh, and Arokiasamy Amit, 2017. "Effective Secure Data Storage in the Cloud by Using the ECC Algorithm", *Middle-East Journal of Scientific Research,* 25, no. 1: 117–127
46. Sun, Alexander, and Bridget Scanlon 2019. "How can big data and machine learning benefit environment and water management: a

survey of methods, applications, and future directions?" *Environmental Research Letters,* 14, no. 7.
47. Taigman, Yaniv, Ming Yang, Marc' Aurelio Ranzato, and Lior Wolf. 2014. "Deepface: Closing the Gap to Human-Level Performance in Face Verification" In *Proceedings of the IEEE Conference on Computer Vision and Pattern Recognition*, 1701–1708
48. Tekerek, A., and Bay, F. 2019. "Design and implementation of an artificial intelligence-based web application firewall model." *Neural Network World,* 29, no. 4: 189–206.
49. Umbaugh, Scott E. (2010). *Digital image processing and analysis: human and computer vision applications with CVIPtools*. Ohio: CRC Press.
50. Wallace, Thomas F., and Michael H. Kremzar 2002. *ERP: Making It Happen: The Implementers' Guide to Success with Enterprise Resource Planning* New Jersey: John Wiley & Sons
51. Wamba-Taguimdje, Serge Lopez, Samuel Fosso, Jean Robert, and Chris Emmanuel TchatchouangWanko 2020. "Influence of artificial intelligence (AI) on firm performance: the business value of AI-based transformation projects" *Business Process Management Journal,* 26, no. 7: 1893–1924.
52. Zhang, Jiawie, and Yu Philip. 2019. *Broad Learning through Fusions*: An Application on Social Networks Switcerland: Springer

CHAPTER 5

BLOCKCHAIN-BASED DIGITAL CURRENCY: SECURING WITH AI AND IOT

SAMIA JAHAN ILMA,
MOHAMMAD SHAMSU UDDIN,
KAMRUN NAHER
ISLAMIC UNIVERSITY OF TECHNOLOGY, BANGLADESH

Abstract

The rising breakthroughs of the Fourth Industrial Revolution (IR4.0)—Internet of Things (IoT) applications, Artificial Intelligence (AI), and blockchain—have significantly altered the business as a whole and unleashed the remarkable potential for virtual currency's future. Blockchain offers a distributed ledger that is both shared and decentralised, may store any asset, and enhances business processes' credibility, transparency, security, and secrecy. The Internet of Things (IoT) improves industrial automation and business process usability by gathering prospective data. AI improves operations by learning from historical data to maximise future results in business processes. Through its immutability, auditability, data security and standardisation, and fault tolerance, blockchain technology has emerged as a promising solution to break the bottlenecks in the financial services ecosystem and ushered in a drastic change from centralised to decentralised financial systems. As a result of the benefits of these inventions and their gradual evolution, the world is gradually becoming a digital utopia with the advent of a cashless society. The triad of these three technologies has resulted in a significant modification of fiat currency and a groundbreaking shift towards digital trading, leading to the emergence of new applications for payment, exchange, and trading infrastructures that are simple, accessible, transparent, and trackable, thereby facilitating financial inclusion. As emerging technologies facilitate the digitization of business models, goods, and services, financial institutions are investing in the development of these innovations to optimise their business processes. Even though technological progress improves the efficiency of financial transactions, the risk of fraud, the assurance of privacy, and the protection of security are perpetual sources of concern. Blockchain technology features openness, trust, veracity, immutability, security, and privacy that help improve IoT data management. AI can lessen the possibility of illegal conduct on the blockchain caused by the anonymity of the transactions by evaluating the data. Therefore, combining these three innovations can optimise their potential for mitigating the cybersecurity threats related to digital trading currencies.

Keywords: blockchain, digital currency, technology integration, artificial intelligence (AI), Internet of Things (IoT)

1. Introduction

Which payment method do you prefer: tapping, waving, inserting, clicking once or twice, or holding out for cold, hard cash? If the second is true, you are about to become the norm. People and companies in every region are finding an increasingly practical use for digital currency. Digital currency is rapidly becoming a viable alternative to traditional fiat currency due to its user-friendliness and high level of security. A growing number of users are moving away from using cash due to the proliferation of cards, the ease and speed of making electronic bank transactions, the rise in popularity of online shopping, and the advent of digital wallets.

As a result of the proliferation of digital and cryptocurrency, the concept of what it means for something to be considered "money" is undergoing a rapid revision process. Because central banks are the ones that issue digital currencies, the banking industry as a whole, as well as how banks make money, is in for some substantial upheaval. In addition, the production of a sizeable quantity of new currencies introduces an element of uncertainty into the functioning of the current monetary system. We have arrived at a pivotal turning point regarding the state that digital money is in right now. A sizeable quantity of physical currency is still in use today, and most governments and organisations still need to transition to establishing a digital currency that a central bank backs. Concurrently, the Bitcoin market is exploding, and an increasing number of people are choosing to forgo the use of cash entirely. More so than at any other time in recent history, governments, financial agencies, global organisations, and banks with decision-making power are investigating digital currencies.

1.1 Opening Case

In an interview with Brink News, Professor Eswar Prasad of Cornell University, Research Associate at the National Bureau of Economic Research, and author of *The Future of Money: How the Digital Revolution Is Transforming Currencies and Finance*, discussed how virtual currency would revolutionise the banking industry. The following section is a paraphrasing of two separate interviews between Professor Eswar Prasad and Brink news, and of his book *The Future of Money*.

In the context of digital currencies, the role that the central bank plays is as follows:

PRASAD: Introducing a central bank digital currency (CBDC) in your country will make competing in the digital payments market significantly more challenging if you work in the financial services industry.

Central banks know a CBDC might limit private sector digital payment processing development. Therefore, they are trying to give their central bank digital currencies as tokens so the private sector can innovate to facilitate payments between customers and businesses.

The business community will likely view this as a very positive development. The increased level of competition in the market for digital payments is expected to result in various benefits, including decreased prices, which will simplify the processing of local and international payments.

Does CBD pose an existential threat to the banking sector?

PRASAD: An account-based CBDC could threaten commercial banks' survival since individuals and businesses may view a central bank account as safer than an interest-bearing deposit account, especially at low interest rates. Central banks know this well.

The Bahamas, which has already adopted its own CBDC, has restricted the amount of money consumers and corporations can keep in their CBDC accounts. These restrictions prevent large sums of money from being withdrawn from company bank accounts.

However, they can no longer rely on some profitable ventures that banks increasingly pursue, particularly those supporting international payments.

In addition, financial technology businesses are beginning to fill the role of a direct middleman between lenders and savers. The core business model of banks is thus seriously at risk from many emerging financial technologies, not simply cryptocurrencies or digital currencies.

Banks will have to compete more aggressively to maintain their current advantage. Even though some banks have already started implementing new technologies, it is possible that the profit margins on banks' core businesses, which include acting as middlemen between savers and borrowers for maturity transformation and payment facilitation, will decline significantly in the future.

Lower Risk for the World of Credit

PRASAD: Regarding money creation, banks will continue to play a pivotal role, as each time a bank grants a loan, it also creates a corresponding deposit. Therefore, commercial banks are primarily responsible for money production in modern nations. If CBDC accounts become widespread, they will compete with commercial banks. Furthermore, this is a significant problem because commercial banks continue to play a crucial role in delivering credit and money to the economy.

We may despise giant banks and not view them as fully secure, but at least we understand how to examine their books and other services.

As more financial activity flows to unregulated or less-regulated sectors of the financial system, we may need to prepare for financial stability issues. This is because regulation has yet to catch up with new technology and all the potential it offers.

CBDC is displacing private cryptocurrency holdings like Facebook's Diem.

PRASAD: Decentralised cryptocurrencies such as Bitcoin or stablecoins like the one Facebook is developing will find it much more difficult to thrive if a digital analogue of a fiat currency is developed.

Diem, Facebook's proposed currency, would be supported by United States dollar reserves, making more efficient transactions within and between nations easier. On the other hand, if a CBDC can perform the same purpose, it needs to be clarified why people would choose a Facebook currency.

There is a possibility that Facebook has particular advantages in innovation, which may make payment systems more effective or user-friendly. However, the justification for a privately generated stablecoin will only be supported if countries like the United States start issuing digital dollars, for instance.

Digital Currency: A Hope or Threat?

PRASAD: We will profit from the incredible financial developments around us in several ways. However, there are also a few issues. Because of the digital test, a private payment provider or the central bank will be aware of the transaction if I buy you a cup of coffee in the future.

People have the right to question whether or not they want to live in a world where each monetary transaction may be viewed by either the government or a private enterprise (like a bank or payment provider). This is because there has already been significant damage done to our privacy. The situation would be considerably worse if every financial transaction could be viewed by either the government or a private corporation.

These new items may threaten the financial system's stability because they are brand new. Because of this, we may need to be mindful of the potential technological and other dangers that lie dormant within them.

Although I have a positive outlook, a cashless future will face significant challenges.

Central Bank Digital Wallet

PRASAD: It might be as simple as loading money into an application, like a prepaid debit card that can be used to make mobile payments. In a more developed form of a central bank digital currency (CBDC), each of us would be provided with an account at the central bank and a digital wallet that we could use to conduct business using digital iterations of the currency. This would allow us to transact with one another using digital forms of currency.

A commercial bank or the country's central bank is the one that manages these digital wallets. In other words, you could go to your bank and have a digital account, just like you do now for your standard interest-bearing deposit account. This would be the same as the current process. On the other hand, you could utilize a digital currency account offered by a central bank for transactions that do not pay interest but allow you to use digital currency for payments in a manner that is not too difficult.

You would require access to these other payment methods, which are the primary differentiator compared to pre-existing electronic payment systems such as debit and credit cards. The government would provide the service to businesses and consumers without charging them.

Change rates and cross-border transactions with different digital currencies

PRASAD: It is predicted that most digital currencies issued by central banks will only be used within their own countries. Given that the vast majority of international payments for commerce and financial activities are currently conducted digitally, the valuation of currencies, also known as exchange rates, will remain the same.

However, as time passes, nations will gradually become more open to allowing their digital currencies to be used more freely beyond their borders. In this scenario, digital iterations of currencies like the dollar will likely be used worldwide. There is also the possibility that digital copies of the Chinese yuan will see widespread use across Asia. On the other hand, the fundamentals of the international monetary system that relate to the flow of capital and currency exchange rates will not be altered.

Transactions on a global scale will become more accessible and less cumbersome. As a result, economic migrants will have an easier time sending remittances back to their home nations, which will also be much easier on their wallets.

Making payments across international borders and tracking them in real time will become a far less complicated process for importers and exporters. As a result, we will see significant improvements in our overall efficiency.

NOTE: How digital currencies will be creating epoch-making transformation in the financial service industry and the era of a cashless society Learn more about CBDC.

Source: Compiled from Brink News (October 4, 2021) "How Will Digital Currencies Change the Financial Sector?", URL: https://www.brinknews.com/how-will-digital-currencies-change-the-financial-sector/;Brinknews (2021, October 3) "The Era of Cash Is Ending," URL: https://www.brinknews.com/the-era-of-cash-is-ending/; *The Future of Money* by Professor Eswar Prasad, Professor at Cornell University and Research Associate, National Bureau of Economic Research, published by Harvard University Press, September 2021, URL: https://www.futureofmoney book.com/the-book/.

There has been a significant digital transformation in the banking and financial services industry in recent years. E-KYC, mobility experiences, and e-trading are becoming increasingly prevalent worldwide due to this trend.

The earlier analysis of Opening Case 1.1 has made it abundantly clear that with the introduction of Central Bank Digital Money, a brand-new form of currency may now be created by utilising cutting-edge technology. This notion was brought up in the previous discussion of Opening Case 1.1.

This low-cost payment system enables customers and clients to send money to retailers via their smartphones without requiring a bank account. Therefore, CBDC can ensure that individuals are receptive to novel ways of utilising financial services.

CBDCs have boosted the public's confidence in blockchain technology compared to their country's banking system. CBDCs can demonstrate such capabilities that central banks can use to demonstrate the resiliency of their payment system. Thus, central bank digital currency may help central banks combat major technology company monopolies.

CBDC can reduce the counterparty credit risk associated with international payments and settlements done by banks in different nations. Everyone paying attention can see how wholesale CBDCs benefit the financial sector and markets. A wholesale CBDC for a jurisdiction that allows cross-border exchange may increase counterparty credit and lower payment and settlement risks.

CBDCs offer a central bank-backed digital currency at a time when digital transactions are becoming more popular. Also declining is the use of automated teller machines for cash withdrawals. Therefore, this is the most efficient way to achieve a cashless society.

1.2 Transforming towards cashless transactions and evolving concerns to protect the process

Central bank digital currencies, also known as CBDCs, offer a competitive advantage over other types of digital currency due to the trust and sense of safety provided by the national central bank. CBDC's long-term viability as an alternative to fiat and digital currency strongly depends on central banks' ability to maximise its benefits.

The creation and distribution of a digital currency could have several advantageous effects. However, security is an essential element that must be considered for all techniques. Similar to physical currency and more conventional electronic payment systems, security precautions for virtual transactions include protections against forgery, fraud, and duplicate spending. The potential impact of a CBDC on the financial and economic systems of a jurisdiction poses various security concerns. Consumer protection and financial stability are also included. Any weak infrastructure supporting digital currency might directly affect consumer payment systems and their users. As a result of attacks or disruptions to CBDC arrangements, the stability of financial markets, economies, and currency-issuing institutions may be jeopardised. A domino effect could trigger these effects.

According to the Reserve Bank of India, the most significant barriers to digital transaction use include cyberspace vulnerabilities and fraudulent online activities.

"The main concern comes from the anglc of cybersecurity and the possibility of digital fraud." We must be cautious about that," said Governor Shaktikanta Das (business-standard.com, n.d.).

Many potential technological solutions have been implemented since then to eradicate this issue. However, security and privacy concerns still haunt us, ultimately affecting the augmentation of digital currency usage. Nationstates will continue to engage in espionage to acquire access to information or destabilise another nation's critical infrastructure. Malicious insiders will continue to use phishing attacks and malware to obtain passwords or private keys in order to steal assets. These two pursuits are likely to continue. Blockchain technology has unlocked a new sector and provided a new glimmer of hope about the cybersecurity of digital currency. Each piece of data saved on the network is authenticated and encrypted, and the network can promptly recognise malicious attacks, ensuring that no data is altered. Blockchain technology offers an open and trustworthy method for recording transactions within the context of a public ledger. Transaction anonymity compromises privacy and control, which can lead to money laundering or terrorism. Privacy demands this sacrifice. AI can reduce crime by analysing blockchain transaction data because of its anonymity. Thus, AI can help prevent crime. Machine learning and profound learning benefit from the IoT's vast data sets. Thus, blockchain, IoT, and AI have greatly benefited digital businesses. Higher data management standards, privacy, and security can usher in a new era of digitization.

Protecting personal information associated with digital currency has become a significant concern in financial services and transactions. Hence, to increase trade security in virtual currencies, the work presented in this chapter focuses on the coupling of blockchain technology and artificial intelligence and their mutual reinforcement through links with Internet of Things devices. In this regard, it has been proposed that the chapter be segmented into sections. The first section will quickly cover blockchain, IoT, and AI's role in digital transformation. Section 2 will provide insights into the blockchain ecosystem for financial services and decentralised finance. Section 3 will discuss the concept of digital currencies and their associated risk measures. The main highlights of Section 4 will be the interconnection between Blockchain, AI, and IoT in improving data standardisation, privacy, security, and scalability. Section 5 will demonstrate the scenario of different banking and financial corporations applying these three innovations to secure virtual currency transactions. Lastly, Section 6 will glimpse future opportunities, scopes, and directions in digital financial trading and end the chapter with a conclusion.

2. Artificial Intelligence and the Internet of Things together with Blockchain: An Ideal Combination

AI, IoT, and blockchain will improve corporate operations and create new business models, enabling digital transformation. Consequently, these technologies will disrupt numerous industries. Experts predict that as many as 28 billion gadgets will be connected to the Internet by 2020. However, just one-third of these gadgets will be personal computers, mobile phones, wearables, and tablet PCs. The remainder will consist of "things" that did not have Internet connectivity earlier. Sensors, terminals, home appliances, thermostats, televisions, automobiles, factory gear, and urban infrastructure are examples of these "things." The Web has transformed into an instant pipeline of enormous information that organisations may examine to improve decision-making, improve performance, and earn revenues due to the accessibility of outgoing data from IoT systems (Perera and Liu 2015). To create and develop an effective AI-powered big data analytics tool, one must first overcome many challenges, including the need for a centralised ecosystem, concerns about cybersecurity, limitations on resource availability, and adequate training data. Blockchain technology provides a decentralised architecture for securing information and functions in an Internet of Things (IoT) ecosystem to eliminate centralised authority and resolve data issues (Atlam, Alenezi, and Alassa, 2020)· Artificial intelligence (AI) can properly analyse data to develop insightful conclusions that benefit the Internet of Things. Blockchain, IoT, and AI are the significant technologies propelling the subsequent digital transformation trend. Combining these three technologies will produce new business models: Revenue will be generated by autonomous sensors, vehicles, machines, and Internet of Things devices. Using IoT, it is possible to create a digital twin. Second, use blockchain to decentralise financial transactions. Third, they independently operate as economic agents using AI and data analytics. Technology fusion accelerates the development of autonomous business models and digital transformation.

2.1 Overview of the Internet of Things (IoT)

If a vacuum cleaner could better understand its user, it might be more akin to a digital mapping system than a traditional vacuum. As a result of strengthening their bond, both the product and the provider have improved. The abbreviation "IoT" for "Internet of Things" no longer serves as a marketing gimmick. The data-driven services they enable are becoming more critical to corporate operations than the actual things themselves. " A system of networked computing devices, mechanical and digital equipment with unique IDs, and the ability to move data over a network without human-to-human or human-to-computer interaction" is the Internet of Things. The Internet of Things can be conceptualised as "a system for linking computing devices, mechanical equipment, and digital equipment," to use the phrase in its most straightforward form. The phrase "a network that connects computing devices and mechanical and digital equipment" accurately describes the "Internet of Things." It is a word that refers to a system where machines and inanimate objects are linked to a network and equipped with sensors, software, network connectivity, and the necessary electronics to gather and share data in real-time. Businesses can learn about you and communicate with you regardless of the technology.

2.1.1 Background of the IoT

MIT professor Kevin Ashton coined the term "Internet of Things" in 1998. His explanation of the functionality of IoT states that it "enables people and things to be connected anytime, anywhere, with anything and anyone, preferably via any path or network and any service" (Sundmaeker and Saint-Exupéry 2010). The adoption of smart home goods such as Amazon Echo, wearables such as Apple Watch and Fitbit, and connected automobiles by AT&T, among others, demonstrates the success of the Internet of Things (Atlam, Walters, and Wills 2018, pp. 928–38). The Internet of Things started with just two computers, but because of the World Wide Web, it has since expanded into a vast network covering a wide range of locations. Mobile Internet is gaining popularity among individuals. Social media use necessitates Internet access. Over time, it evolved into the Internet of Things, a world of connected objects (Perera, Liu, and Jayawardena 2015, pp. 1–13). In 2008, there were more connected gadgets than people living on Earth. According to projections by Statista, there will likely be more than 75 billion IoT devices in operation around the globe by the year 2025.

The benefits of an internet-connected "Internet of Things" go well beyond internet-connected consumer devices. The BFSI (banking, financial services, and insurance) industries may undergo significant change due to the growth of linked devices. Due to the vast amount of data they are responsible for gathering and distributing, financial institutions, such as banks, can tremendously benefit from the IoT's assistance in enhancing and streamlining their operations. In addition, the IoT has the potential to enhance security, reduce risk, and increase customer satisfaction for financial institutions by introducing digital payment systems that allow for transactions without bank involvement and self-service customer support chatbots (Som&Kayal 2020, pp. 141–61).

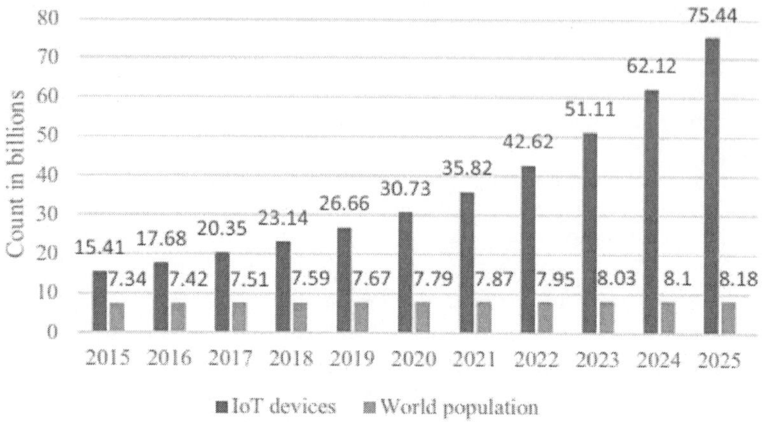

Figure 1: By 2025, each individual will rely on roughly ten linked devices, according to research forecasts. Data Source: Statista

2.1.2 IoT Digitising the Business Landscape

How IoT works in a company: The digital transformation and the Internet of Things rely mainly on data to reach their full potential. The Internet of Things allows businesses to utilise client data directly. It enables businesses to be physically present in the same room as their customers. When deployed appropriately, IoT technology can explore the physical world for data sources and convert them into valuable business insights. The source of information from the inside of a device begins at the sensor. For the data to be transmitted over the network, it must first be converted into a digital payload by a software agent running on the embedded system. Then it must be encapsulated by protocols. This node, sometimes called the edge of the IoT network, collects data payloads before delivering them to the IT network via the OT network. After being stored in a database, the data is made available on the cloud. Next, analytical tools, or AI, can be used to sift through the data and draw conclusions. Models are created through this processing, and the data they produce is saved in the company's back-end databases. This information is shared with other departments so they can use it to standardise and improve their procedures.

Scope of companies through IoT platforms: Businesses now highly value data as AI has made it possible to act on it. Advances in artificial intelligence have allowed for the proliferation of digital IoT applications, enabling predictive and prospective analytics. Experts agree that by 2025, more than 80 billion gadgets will be connected to the Internet, fuelling the IoT's rapid expansion. IDC has released a report investigating the revolutionary changes brought about by digital transformation in modern businesses. According to Figure2 from the report "Internet of Things and Digital Transformation: An Account of Four Industries," By 2018, 33% of global industries will be disrupted, and digital transformation is an imperative necessity to protect your business from the dangers posed by digitally savvy competition.

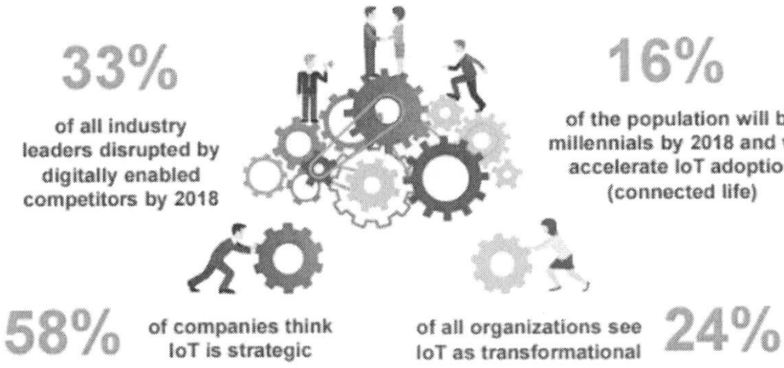

Figure 2: How digital transformation is revolutionising modern enterprises Source: 2015 August Global IoT Decision-Maker Survey by IDC

It also demonstrates why MNCs see IoT as a strategic instrument for growing their businesses' scalability and how IoT is helping them develop fresh, exciting new ways to interact with their customers. Creating novel approaches to better serve customers is the first phase of digital transformation. Businesses need help shifting from traditional methods to mass-market strategies driven almost entirely by consumer needs. This is made possible by the Internet of Things, which connects customers and businesses regardless of the device they are using and delivers the needed information to customers. Connecting gadgets is another way the Internet of Things has simplified the online acquisition of goods and services. When devices are connected, it is simple for buyers to review a product on their phone while they are out and then on their laptop at home. Customers have an insatiable thirst for a personalised and pleasurable experience from the businesses they tackle. Customers want to be able to interact with sellers whenever it is convenient for them; they want an abundance of options to choose from that are tailored to meet the particular requirements of their lives; and they want in-depth information about the products they find most interesting. They want the purchasing process to be streamlined. Customers desire prompt service and an atmosphere without potential dangers when purchasing.

How IoT is reforming business models and helping companies gain a competitive edge: Businesses must create a whole new digital infrastructure to meet the networked economy's needs and expand customer services. This will significantly impact how businesses operate and open new value generation and capture options. The technology stack is depicted in Figure 3 from the Harvard Business Review. It consists of new hardware, software applications, a network connection, the "product cloud" hosted on remote servers, security tools, a doorway for external sources of information, and the incorporation of corporate business systems. The technological stack includes each of these components. Businesses that use cutting-edge technology as soon as it becomes available are more likely to succeed. Take, for instance, the widely used Uber service. Despite not owning any taxis, Uber has successfully disrupted the taxi industry in many countries. It had developed into a multibillion-dollar company three years after its founding. Similarly, it is projected that these businesses will be able to reach that size in a single year in the not-too-distant future. It is given in Figure 3.

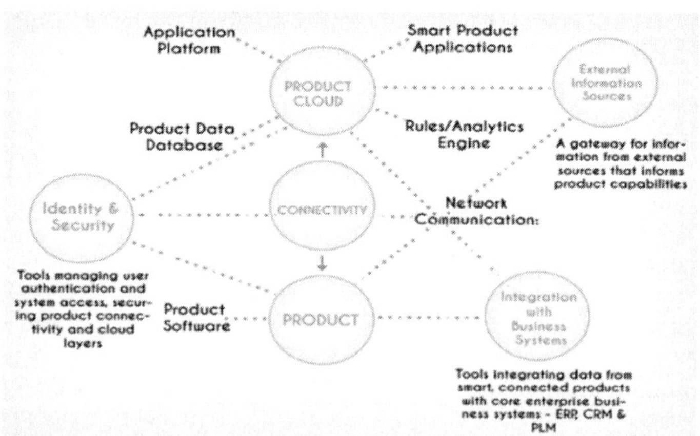

Figure 3: Impact of IoT on business models
(Ref.: https://hbr.org/data-visuals/2015/09/the-new-technology-stack)

Smart and connected goods symbolise a new generation of technological capabilities that can provide firms with a significant competitive edge by providing enormous quantities of diverse data in real-time. This purpose can be achieved by utilising the Internet of Things (IoT). Productivity will increase significantly due to these things' ability to update their owners' businesses in real time, regardless of where or when they are used. This results in a broader understanding of the product's utilisation and may yield priceless insights. It is possible to make better engineering and design decisions based on data-driven insights on the performance of specific commodities and the ecosystem.

This feedback will be employed to improve value chain governance initiatives, resulting in modifications to business model architectures. The benefits of digital transformation can only be achieved by enhancing engineering, marketing, sales, and customer support close to the customer. The IoT has been significant in assisting businesses to become more connected and enhancing consumers' shopping experiences over the past decade or so.

Overview of Artificial Intelligence (AI)

2.2.1 AI Define

Programming computers to do human jobs is called artificial intelligence (AI). AI is an attempt to give computers the ability to think like humans. It can be thought of as thinking, reasoning, and solving problems, or as a collection of techniques that enable computers to programme themselves. The Study of Artificial Intelligence Integrates Numerous Disciplines, Including Natural Language Processing, Computer Vision, the Internet of Things, and Robotics, with an Emphasis on the Integration of Statistics and Machine Learning Techniques (Som and Kayal 2020, pp. 141-61).

AI can be viewed as a fully functional organism against which its relationship to other subfields can be evaluated. Machine learning is the system-wide development and customization of specific actions, activities, or thoughts. The Internet of Things depends on people's perception and interpretation of their surroundings. Lastly, extensive data can be compared to the food and oxygen that provide energy and allow the system to function. This is a somewhat simple illustration, but it explains the relationship between all the words. The decentralised and trustworthy nature of blockchain technology, together with the immutability and transparency of its cryptographic trail, enables the secure transmission of artificial intelligence data, techniques, and models without compromising their legitimacy. The distributed ledger technology that is blockchain has the potential to make business activities more trustworthy, accessible, dependable, and private. It provides a shared, decentralised, and distributed ledger. At the same time, AI can improve operations by seeing trends and determining how to achieve the best outcomes (Som and Kayal 2020, pp. 141–61).

2.2.2 AI Relies on data

The failure of digital transformations is mainly attributable to businesses' need for more use of artificial intelligence.

Customers expect enhanced and expedited experiences without having the patience to wait. Customers compare their purchasing experiences with your company to those of industry leaders. Emerging revolutions generate new approaches for offering value, creating distinctive customer experiences, and keeping up with consumers' pace. It represents a substantial shift in corporate operations. A data-driven digital transformation necessitates artificial intelligence. The only way to be genuinely productive and provide enterprise-wide insights to make sense of all this information is to develop effective and comprehensive data literacy. Establishing systems and resources capable of integrating, organising, and analysing this data is vital. Companies recognise the significance of investing in novel technology and procedures capable of evaluating data more rapidly and smoothly, enabling them to grasp the advantages of these expenditures. Moreover, the only way to unlock the potential of this information in the present day is to adopt an AI transformation strategy.

2.2.3 AI: the key driver of digital transformation

AI is a stimulant for digital transformation, enabling businesses to be more inventive and data-driven and to cope with consumers' rising pace, resulting in beneficial and lucrative transformational outcomes (Pflaum and Michahelles 2018). According to a recent Infosys survey, businesses that use AI to propel their digital transformation produce 15% more revenue, with 98% of these businesses reporting this result.

Enhance the Customer Experience: A successful digital transformation approach relies on accurate customer profiles and audience identification. AI contributes to creating a comprehensive, 360-degree image of the consumer by revealing patterns in buying behaviour, past purchases, and digital channel involvement. In the current digital realm, consumers use multiple channels and a growing number of interactions during the buying cycle. The distinction between virtual and in-store shopping is fading. Consumers anticipate a seamless experience between channels and hyper-relevant facts and messages from businesses when they engage. Along the buyer's journey, consumers anticipate increasingly customised experiences. AI is a crucial tool for rapidly uncovering customer information that can then be used to build customised consumer experiences. The more usable consumer data a company possesses, the more service it can provide in those all-important micro-moments.

Robust Analysis of Customer Data: AI can process thousands of data pieces in real-time to provide insights and find trends. An online shopper, for instance, can browse an e-commerce website and obtain individualised offers based on browsing data, previous interests, and buying activity. Dynamic analytics backed by artificial intelligence enable firms to anticipate consumer requirements rather than respond to consumer activities. Brands with a fully functional digital ecosystem may provide relevant messages, information, and product recommendations.

Predictive analytics uses AI to combine insights and patterns from different data points and algorithms, including purchase history, promotional information, and consumption patterns. Based on the insights, marketers can design highly tailored consumer experiences and make data-driven decisions to develop more efficient and profitable campaigns, saving money on budgeting, media, and advertising.

Increase the effectiveness of business processes: procedures may be expanded quickly and provide reliable data to accelerate decision-making. Trends and patterns help make risk and opportunity judgements. This enables the early detection of bottlenecks. Advertising expenditures can be better allocated to maximise returns and KPIs. AI can propose the best plan of action for optimal results. AI can improve the customer experience with chatbots and data analytics. Amazon uses AI to administer customer support and determine when clients need live representatives.

2.2.4 Scaling up business operations with AI

Today's CIOs and IT leaders perceive digital transformation as a vital endeavour instead of an internal productivity-enhancing plan. Recent developments have driven businesses to adopt digitization, resulting in widespread acceptance of the Fourth Industrial Revolution and technologies like artificial intelligence. Gartner finds that just 53% of AI projects get from concept to production, but businesses must see the value of AI adoption. AIoT, a combination of sophisticated AI and the Internet of Things, possesses characteristics that can be advantageous when both technologies are employed. The Internet of Things enables the intelligent optimisation and real-time coordination of physical and digital processes across process control systems (PCS), manufacturing execution systems (MES), enterprise resource planning (ERP), and other technologies to increase their combined effectiveness. Specifically, these capabilities aim to improve the efficiency of these technologies. IVR is a market-leading AI solution due to its ability to process enormous volumes of data. Businesses can utilise conversational AI to enhance customer experience, IVR containment, and omnichannel collaboration to optimise the potential for cross-selling and upselling, thereby gaining control of the entire platform and overcoming problems resulting from a lack of conversational context. Consequently, artificial intelligence offers increased speed, convenience, and cost-effectiveness while simplifying complex processes and systems.

Blockchain

The exponential growth of blockchain technology over the past couple of decades has proven many previously held beliefs untrue, including the impossibility of eliminating transaction costs, preventing double spending, eliminating net fraud, recovering lost data, and so on.

2.3.1 The Blockchain Concept

Blockchain is a distributed, immutable record that tracks P2P transactions. In the ledger, each transaction list is contained within its own block. Each block consists of a log file with a date and a hash algorithm connecting it to other blocks in the chain. In the end, what will be constructed is a chain of blocks, often known as blockchains. It is aptly described as "a decentralised database consisting of sequential blocks of digitally signed asset transactions that are cryptographically linked and controlled by a consensus technique" (Sultan, Ruhi, and Lakhani 2018, pp. 49–57). Most blockchain network stakeholders must agree to record a transaction in the public ledger. By ensuring that each node that contributes to the blockchain obtains a copy of the legitimate ledger, distributed ledger technology makes it simpler to communicate information. This ensures that all users are updated on newly added blocks or swaps. The blockchain database's records are accessible and verifiable due to their distribution. In a central repository, this data cannot be tampered with. Its data is hosted online by millions of machines.

2.3.2: How Blockchain Works

Users have public and private keys. These two secured cryptographic keys allow restricted system interaction. Say two people exchange bitcoin. A transaction can be initiated using either the public key or the user's private key. Other users can accept the transaction by using their keys. Together, they can use the public P2P system to transact. Currently, the system checks the transaction's details to ensure consistency with previous data. The transaction is refused if these nodes concur that the originating user does not have the claimed coin. If additional nodes agree, the transaction becomes a "block" in the chain (Goel 2018). It is depicted in Figure 4.

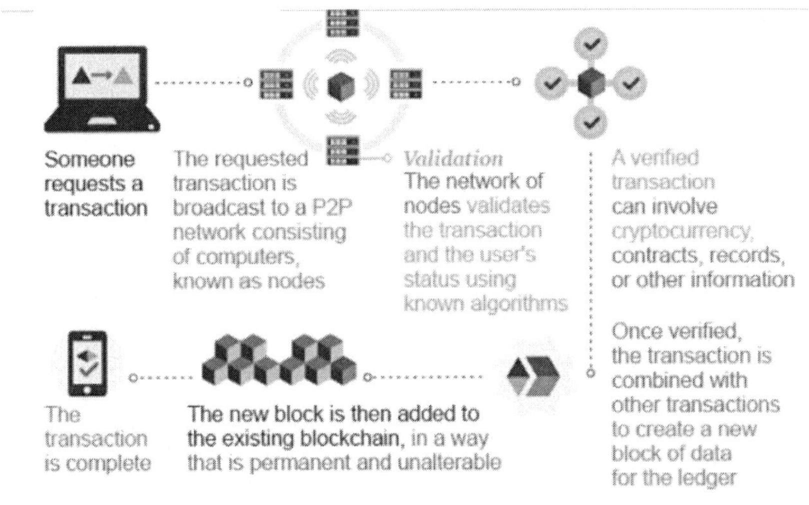

Figure 4: The way blockchain technology facilitates a transaction

2.3.3 Blockchain: A security guard of data

The rise in identity theft is a drawback of the digital transformation. Cybercriminals aim their attacks at government databases. Data breaches, such as Equifax's in 2017, have exposed the personal information of millions of Americans. According to Booz Allen Hamilton, blockchain data topologies strengthen network security by lowering the probability of a database compromise due to a single point of failure. According to

McKinsey, governments could provide a trustworthy electronic version of citizen documentation by storing their hashes on the blockchain.

The following are the main features that make blockchain technology so secure:

i) A ledger is utilized by blockchain technology to record all financial transactions. This type of "master" ledger would ordinarily be an obvious risk point. If the ledger is corrupted, it might cause the system to fail. For instance, if someone edited a record, they could steal enormous amounts of money. Alternatively, they could acquire access to confidential personal information if they read all the transactions. The ledger on the blockchain is decentralised. This indicates that no single computer or system controls the ledger at any given time. This level of access to the central ledger would require an exact, large-scale attack on thousands of devices simultaneously to access the authentic ledger.

ii) The chain itself is another crucial security feature. The ledger is a long string of blocks linked together in the correct order. The interconnectedness of these chains is significant. These records span from the beginning of the system's creation until now. This means that if someone wanted to amend a transaction, they would have to change every transaction in the chain leading up to the one they wanted to change. This significantly increases the difficulty of the imagined tampering process. This also significantly improves the system's general safety.

iii) More than two users must concur that a transaction is legitimate for it to go through. The number of steps in the verification procedure is relatively small, even in the most advanced current payment processing systems. There is a purchaser, a vendor, and possibly some more parties (a bank or credit agency). Hundreds, if not thousands, of individual nodes make up a blockchain network. The digital ledger is replicated in its entirety on every node. Each of these can be used on its own to confirm the transaction. If some of the nodes in the network do not approve of the transaction, it will not go through. The ledger is kept organised thanks to this approach. Furthermore, it renders fraudulent transactions extremely difficult.

iv) The cryptographic keys utilised in blockchain exchanges are also technological marvels. Each cryptographic sequence is lengthy, intricate, and nearly impossible to interpret without the proper authorization. The two-key approach increases the security of transactions. Furthermore, it does so without jeopardising the unprecedented transparency of the system (Goel 2018).

2.3.4 Blockchain: Diminishing the Gaps of IoT

In the beginning, cryptocurrencies and financial transactions were the primary uses of blockchain technology. All nodes in the network were responsible for processing and recording transactions. Consequently, blockchain is already being implemented in various industries due to its numerous advantages. The Internet of Things (IoT) is a hypothetical example. Combining blockchain and IoT can offer several benefits (Atlam and Wills, 2020). Combining IoT and blockchain provides tremendous benefits. Using the shared system features of blockchains, for instance, can reduce instabilities and a single point of failure in a centralised IoT ecosystem by eliminating the need for a database controller to manage IoT sensors and their networking services with one another. Blockchain's powerful cryptographic algorithms, hash functions, and time stamps boost cloud infrastructure security and privacy. Blockchain technology offers an immutable and tamper-proof ledger to protect data from hostile threats. Any record modification can only be cached in the ledger if many attributing users confirm it (Karafiloski and Mishev 2017, pp. 763–68). Therefore, a trustworthy system is built in which the participating IoT applications appear to be the only entities capable of approving or rejecting a transaction based on their consent (Reyna et al. 2018, pp. 173–90). The advantages of integrating blockchain technology with the Internet of Things are illustrated in Figure 5.

The most crucial step is choosing the blockchain platform to link IoT to blockchain. Since Ethereum, IoTA, and Hyperledger are all open-source and permissionless systems, they are now the three most widely used protocols. In addition, they may include block linking, a hashing of transaction data within a single block that is cryptographically protected, innovative trails, consensus, and smart contracts. A wide variety of industries, including banking, mortgage lending, mass transit, internet broadcasting, and healthcare, utilise blockchain technology for data and creditworthiness protection. Blockchain technology can be used for much more than just Bitcoin, although the general public tends to associate it almost exclusively with Bitcoin. Integrating the Internet of Things with blockchain technology is necessary for supply chain and logistics operations. Monitoring from beginning to end is made more difficult by the presence of various partners along a global value chain. When combined, blockchain and the Internet of Things have the potential to dramatically increase dependability and provenance. The Internet of Things may capture data on movement, climate, location, and other variables, which will then be stored on the distributed ledger.

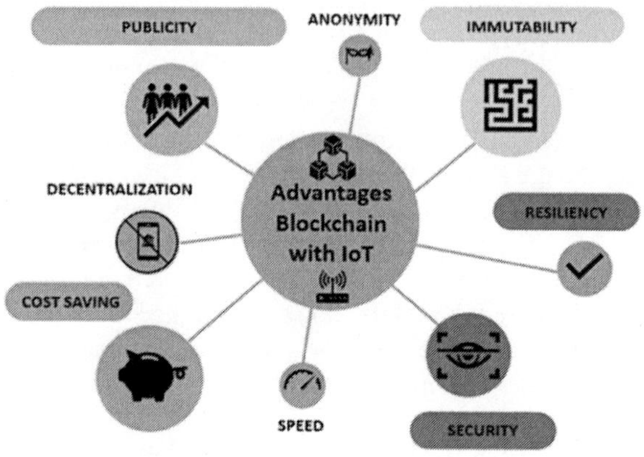

Figure 5: Combining Blockchain with IoT Benefits (Source: Ankit Som and ParthajitKayal, who have written extensively on AI, blockchain, and IoT, are responsible for compiling this work.)

2.3.5 Power of Blockchain: Accelerator of Digitization

New shifts and changes in recent years have surrounded the corporate sector. In an era of pervasive digital transformation, businesses must make strategic decisions regarding arranging and integrating company operations into functional business models to thrive. For this reason, the management team must grasp the innovative possibilities digital technologies provide for their businesses without delay. COVID-19 has functioned as an accelerator, compelling businesses worldwide to implement creative digital solutions to address the pandemic's effects. Organisations must comprehend the potential for emerging technologies such as blockchain to contribute value to digital transformation strategies as we adapt to hybrid work modes. Database technology's fundamental characteristics were outlined in 1991, but it was not until 2009 that blockchain was first put into action with the digital currency Bitcoin. Since then, the potential applications of blockchain have expanded rapidly. However, the innovation is still in its infancy in many locations. Despite this, experts are confident that blockchain will profoundly affect the business world in the following years.

Digital Transaction:

The decentralised framework, in which network participants have a database prototype, and cryptographic technologies ensure that no user can change database information surreptitiously. Each transaction must be sent to the system as a cryptographically secured block and evaluated. Because there is no need for a third party in peer-to-peer transactions, blockchain naturally creates high credibility. Blockchain technology may become an essential digitization tool as business-to-business and business-to-customer transactions increasingly occur online. It enhances confidence and security for consumers, companies, and trade and business partners in anonymous cross-border digital communication. As blockchain is well recognised as the foundation for Bitcoin and other cryptocurrencies, industry leaders are quickly adopting it. Fortune 500 financial services and technology companies are testing blockchain for international financial transactions. JPMorgan Chase built the JPM Coin network to transmit cash-backed cryptocurrencies across borders.

Business Processes:

Speed is a crucial differentiator in digital transformation. Moreover, storing and transferring data on a blockchain offers excellent security. On this foundation, you can automate commercial transactions. Manufacturing uses blockchain technology to track components and raw materials through the manufacturing cycle, making it possible to track components back to their original manufacturer in case of a recall or malfunction. For example, carriers, customs, freight forwarders, and port authorities have adopted TradeLens, a blockchain network for international shipments. This has increased transparency for all parties involved in the shipping process while reducing transit times and maximising efficiency.

Supply Chain:

The blockchain's decentralised structure ensures complete visibility throughout the supply chain, from sourcing components to selling finished goods. Simple and cheap transaction recording on the blockchain is possible. They are lessening the likelihood of problems and making it easier for the final consumer to know precisely where their goods came from. For its immutable audit trail, it has been used by the primary food and beverage industries, international transport firms, and commodity distributors.

Walmart's produce suppliers must join its Food Trust blockchain network. Blockchain DLT offers end-to-end tracking of veggies to their farm of origin. What is the significance? In 2018 and 2029, romaine lettuce caused E. coli outbreaks. Grocers were compelled to remove all lettuce from stores since they could not trace the tainted lettuce to specific producers. The blockchain alters that. Lettuce packages can be dispatched using secure QR codes connected with a blockchain-based tracking app. Grocers and consumers can track lettuce from farm to shelf using a blockchain-based, immutable ledger.

3. Blockchain ecosystem for financial services and decentralised finance

Every day, more people who support implementing distributed ledger technology (DLT) are coming forward in the financial services (FS) sector. Why not, then? Blockchain and distributed ledger technology (DLT) have demonstrated considerable promise. In the financial sector, the emergence of revolutionary technology such as blockchain has produced tremendous tension among market leaders. Leading financial institutions are experiencing increasing discomfort with the entry of new high-tech competitors into niches that they have traditionally dominated for centuries. The advent of blockchain as a new high-tech ecosystem may impact the financial industry's institutional nature, financial transactions, trading patterns, and participant relationships (Barbu, Militaru, and Ionut 2020, pp. 44–57).

It promises significant infrastructure, transaction, and administrative cost benefits for financial institutions (FIs). It can eliminate the need for central counterparties in the digital transfer of financial assets. Additionally, it can support the financial ecosystem's stability, precision, and dependability. By 2022, blockchain may save banks US$20 billion in infrastructure expenses for securities trading, foreign payments, and regulatory compliance, according to a 2015 Santander estimate ("Perspective Blockchain Adoption in," n.d.).

Blockchain has become a valuable tool for getting around bottlenecks in financial services because it is decentralised, cannot be changed, can be audited, and can handle mistakes. Table 1 describes the features of blockchain technology that contribute to its diverse uses in the financial services industry.

Table 1: Attributes of Blockchain Technology in the Financial Service Sector.

Critical features of Blockchain Technology facilitating the Financial Service ecosystem
Contains a distributed network of parties' shared secure, computerized, time-stamped transaction ledger database.
It offers a precise and traceable record of each transaction, as well as a transaction history that can be viewed but not edited, and this history can also be viewed.
It is built on a P2P cryptographic network that verifies existence, functionality, and origin.
Consensus and triple-entry accounting identify who owns securities, virtual money, and other assets.
Has a massive global computer network-verified transaction ledger—identical ledger copies on different platforms managed by different organizations.
Enables users to examine the blockchain records. However, users can only update the Blockchain with the majority's consent.
Ledgers have two possible settings: public or private and permissioned or not permissioned.
Can establish transaction-specific business rules regarding a transaction. Enables smart contracts, which can be automatically executed and have their terms written in computer language.
Uses encryption to provide immutability and irreversibility; information, once input, cannot be deleted.
Specifics of transactions are recorded without disclosing sensitive information about the parties or subjects concerned. Enables settlement in almost real-time.

Source: infosys.com

3.1 Financial Service Sectors for Applying Blockchain

Blockchain technology can improve regulatory efficiency (real-time financial activity monitoring), facilitate and simplify operations, reduce counterparty risks (execution of agreements in a shared and immutable environment), disintermediate settlement and clearing of transactions, and transact.

Three well-documented and dynamic subsectors can be used to generalise the wide range of use cases. Even in emerging markets, use cases are being examined, and market proof has been completed or is being completed in various subsectors.

Commercial finance: products and service transactions demand end-to-end transparency in commercial financing. Blockchain technology facilitates suppliers' use of diverse and incompatible systems by showing transactions to all users on a distributed ledger. Blockchain can give visibility into the order-to-delivery pipeline, reduce the number of inconsistencies, and reduce the time needed to address disputes.

International transactions: To ease foreign currency operations, banks need a more straightforward reconciliation approach to manage Nostro and Vostro accounts, where Nostro accounts are local and Vostro accounts are worldwide. By automating the reconciliation of all accounts, accounts can be recorded on a blockchain as a transaction, enhancing efficiency and transparency. Due to the ability to manage all bank transactions through a single interface, it is now possible to see transaction progress, currency balance, process health, consistency, and timeliness more clearly.

Trade financing: It calls for accelerating the procedure for obtaining approvals from various governmental agencies, including customs, port authorities, transportation authorities, and more. Blockchain technology enables legal trade companies to sign approvals, communicate status updates, and update payment information. Blockchain technology simplifies complex procedures by allowing parties to access a shared ledger. The blockchain system's elimination of drawn-out settlement and error dispute timeframes reduces the time needed to access funds. This approach makes businesses, thc government, and the public accountable. Figure 10 demonstrates the blockchain-enabled trade financing cycle (Anderson 2019, pp. 349–56; Son et al. 2019).

Blockchain Technology Escalating Trade Finance:

Instant financial document approval is linked and accessible over DLT, which helps reduce the time needed to initiate a shipment.

Transparency in factoring: Accessing invoices through distributed ledger technology (DLT) allows for real-time transparency on short-term borrowing.

Ownership Authentication: The availability of title within DLT provides transparency into the ownership and whereabouts of items.

Legislative and regulatory neutrality: Regulators have real-time access to critical records to aid in AML and enforcement actions.

Bills of lading are tracked using DLT, which reduces counterparty risk and prevents double expenditure.

Decentralised contracting: When the contract terms are met, the distributed ledger (DL) is updated in real-time, an additional benefit of adopting blockchain technology in the financial services business. This makes transactions more efficient and streamlined by reducing the time and resources required to supervise the delivery of items and services.

Cost savings and automated settlement: Additionally, adopting smart contracts on blockchain networks can result in reduced transaction fees and accelerated, automated settlement. By eliminating the need for intermediaries, such as correspondent banks, the processing of transactions becomes more cost-effective and efficient.

Eliminating intermediaries: By facilitating trade finance using DLT, banks eliminate the requirement for a trusted intermediary. Correspondent banks are no longer required ("Perspective Blockchain Adoption in," n.d.).

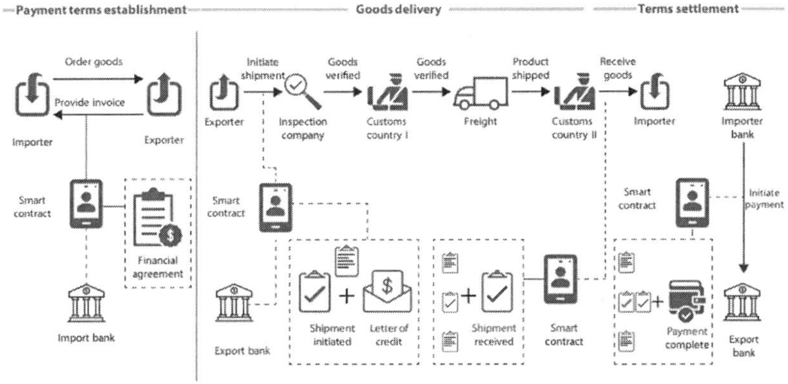

Figure 6: Trade finance process flow under blockchain setup. Source: infosys.com

3.2 Blockchain in FS business applications

Since 2009, blockchain technology has expanded beyond Bitcoin applications to all FS industry sectors. Consider the following probable FS business domains:

Commercial Banking:

- Products denominated in cryptocurrencies (e.g., from Tinker, SolidX).
- Tracking assets and properties; registering tangible assets (house, land, automobile)
- Market, peer-to-peer, and syndicated lending. Smart contracts are used for mechanical servicing and real-time loan funding.
- Smart contracts for funding loans in real-time and automated service of those loans
- Financial management on an individual level (PFM)
- Liquidity, cash reserve, and intra-bank settlements
- Management of customer acquisition and loyalty.

Finance for trade and supply chains:

- Real-time multiparty tracking and control of instruments used for open accounts, bank payments, and letters of credit.
- Receivable financing.
- Commodities trade finance
- Decentralised contract management, factoring, and debt servicing
- Documentation services (trusted private e-doc exchange, real-time review, and approval of documents)
- Export-import banks (eliminating the role of correspondent banks)

Capital Markets:

- Clearance and settlement (Hyperledger, Serica).
- Completed trade (real-time transaction matching, automated DVP on the cash ledger).
- Post-trade (trade reconciliation, trade reporting, monitoring, and surveillance).
- Services for security and custody (escrow and custodian services, asset documentation, record-keeping).
- A trade involving derivatives.
- Asset registration, maintenance, documentation, and transaction.

Payments:

- Retail payments and micropayments.

- Large-scale agreements (correspondent banking networks, cross-border FX).
- Peer-to-peer transactions (BitBond, BTC Jam).
- Processing payments (e.g., Coinbase, BitPay).
- Offers for digital wallets and exchanges (e.g., Bitreserve, BitPesa).
- Currency conversion and international money transfers (MeXBT, Coinbase [Wallet], Kraken, and Ripple).

Regulatory Compliance:

- Automate compliance-related duties' execution (e.g., real-time regulatory control limit enforcement [e.g., for asset rehypothecation], CCAR-related).
- Regulation optimisation (for example, in CDD, AML, and KYC).
- Sanctions enforcement; regulatory tools (e.g., for parsing real-time feeds from FIs; audit trails for compliance verification).
- Automatic reporting to regulators (through smart contracts, DL as a golden source, and unified regulatory reporting protocols).

Risk Management:

- Risk assessment and risk management.
- Managing counterparty risk.
- Fraud risk management and identity theft prevention.
- Capital risk management and liquidity risk management.
- lowering of systemic risk (real-time global view), lowering of operational risk ("Perspective Blockchain Adoption in," n.d.).

3.3 Decentralised Finance (DeFi): An umbrella phrase for financial applications

Cryptocurrency is an umbrella term for various decentralised digital currencies and payment methods. A new financial ecosystem has been created by utilising blockchain technology, which is characterised by being decentralised, distributed, immutable, and self-governing. DeFi facilitates peer-to-peer finance through the Ethereum blockchain, building a network of interconnected financial protocols and solutions such as borrowing and lending platforms, stablecoins, and tokenized bitcoin. It is now the most active sector in the blockchain industry, with over $13 billion invested in Ethereum smart contracts and numerous opportunities for people, developers, and businesses.

Unlike the old financial system, the Ethereum blockchain is the foundation of decentralized finance, which relies on a centralized infrastructure maintained by authorities, institutions, and intermediaries. Consider two individuals: one who wants to lend some surplus money, and the other who needs to borrow funds to build their business. In the past, they needed to interact with a centralised financial organisation like a bank. This was one of the requirements. Additionally, the bank can assess if a customer is eligible for these services. With DeFi, however, there is no need for dialogue; all the lender and borrower require is interaction with a smart contract outlining the loan's terms and conditions.

Apps that run independently from a central server are the foundation of DeFi. These protocols that enable consumers to use DeFi's full potential are crucial because they free up smart contracts to undertake the heavy lifting in place of a centralised financial institution. Existing public blockchains, such as Cronos or Ethereum, may be the foundation for DAPPs. All nodes in a blockchain must agree on any changes before they can be made. Once a status or piece of information has been posted to the network, it cannot be removed or altered in any way. Due to this openness, decentralised applications (DAPPs) gain the advantage of being censorship-resistant and self-sufficient. A decentralised lending platform can lend stable coins pegged to the US dollar (USD). Participation in DeFi and putting coins in a decentralised liquidity pool or on-chain investment fund make it possible to earn interest on cryptocurrencies. This illustrates De-Fi (Schär 2021, pp. 153–74).

3.3.1 Smart Contracts: The Backbone of De-Fi

Smart contracts are sometimes viewed as simple programs simultaneously performed and saved by several validators on a blockchain. The execution of a smart contract may be depended upon, and any interested party can independently validate the subsequent state changes. Properly executed smart contracts provide high transparency and reduce manipulation and arbitrary meddling.

In order to fully grasp the revolutionary nature of smart contracts, we must first become familiar with conventional server-based web apps. Users who interact with this programme should be familiar with its underlying mechanisms. As a further difficulty, the user is prohibited from modifying the running conditions. It would be feasible to manipulate one or both of the variables. Therefore, the user is required to have faith in the ASP. These concerns can be prevented or corrected by smart contracts, ensuring that an application will operate as planned. Since their code is stored on the blockchain and can be freely viewed by anyone, smart contracts are transparent. Thousands of network participants also carry them out via transactions that activate its functionality. These executions result in modifications to the contract's state, such as an update to the account balance, which are recorded on the blockchain network via the state tree. This secures the legitimacy of the execution and the predetermined behaviour of the contract by adhering to the blockchain's consensus rules.

Smart contracts are incredibly versatile, given the various orders they may execute. They can also operate as custodians, storing Bitcoin assets and releasing them according to predetermined criteria. These characteristics allow for robust ecosystems and grant countless inventive opportunities. Szabo initially proposed the notion of smart contracts in 1994 and refined it further in 1997, comparing it to a vending machine and explaining that various agreements might be integrated into the hardware and software we interact with to make contract violations costly for the perpetrator. Buterin subsequently proposed a blockchain-based architecture for smart contracts to address execution environment problems and enable safe global states. Additionally, this architecture permits contracts to interact and build upon one another (Buterin 2009, pp. 1–36). Wood enhanced the concept now known as Ethereum, the most widely deployed smart contract platform. Despite the availability of various other options, Ethereum remains the most popular option (Wood 2014, pp. 1–32).

3.3.2 The De-Fi Infrastructure

The DeFi system comprises numerous interconnected levels, each with a distinct function. The interdependence of the layers results in a flexible, open infrastructure that anyone can access. It is essential to recognise the hierarchical nature of these layers, as the security of higher layers depends on the integrity of lower ones. For instance, if the blockchain in the settlement layer is corrupted, it might expose all upper levels to risk. If the initial layer employs a permissioned ledger, obtaining a higher degree of decentralisation becomes difficult. Figure 9 illustrates a theoretical framework for examining these layers, namely the token and protocol layers. It describes the five distinct layers: settlement, asset, protocol, application, and aggregation. These are known as protocol layers.

The blockchain and protocol assets comprise Layer 1 of the architecture, the settlement layer (e.g., Bitcoin [BTC] on the Bitcoin blockchain and ETH on the Ethereum blockchain). It secures ownership information and ensures that system updates follow its regulations. For trustless execution to occur, the blockchain must serve as a settlement and dispute resolution layer.

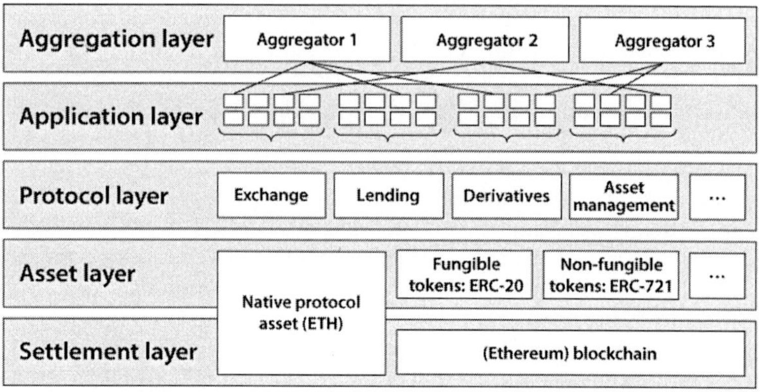

Figure 7: The building blocks of De-Fi infrastructure Source: The DeFi Stack

1. The second layer contains all assets issued following the settlement layer (the asset layer). The term "asset" refers to both the asset for the native protocol and any other assets published on this network (usually referred to as tokens). Interest on coins can be earned by depositing them in a decentralized liquidity pool or an on-chain investment fund.
2. Thirdly, the protocol layer (Layer 3) defines the rules for specific applications, including on-chain asset management, debt markets, derivatives, and off-chain markets. Most often, accessible smart contracts are used to implement these criteria (or DeFi applications). As a result, specific protocols can be used with others.
3. Application layers (Layer 4) design user-centric applications that utilise particular protocols. Web browser-based interfaces simplify smart contract interaction.
4. Adding an extra aggregation layer on top of the application layer is what happens in Layer 5. Aggregators build user-focused systems with support for several protocols and programmes. Common features include rating and comparison tools for various services,

connecting to multiple protocols simultaneously to complete otherwise tricky activities, and consolidating all essential data into one easy-to-understand place (Schär 2021, 153–74).

3.3.3 Advantages of Decentralised Finance

- **Programmability:** The high programmability of smart contracts enables the automation of the production of new financial instruments and digital assets.
- **Security and privacy** Due to its distributed nature, the blockchain enhances security and audibility by coordinating data in an immutable manner.
- **Connectivity:** Ethereum's modular software architecture enables smooth integration with DeFi protocols and applications. Software developers and product managers are free to work with any protocol, design their user interfaces, and interact with others, thanks to DeFi. Therefore, DeFi protocols are also known as "money legos."
- **Transparency:** Other users broadcast and validate transactions on the Ethereum public blockchain. Addresses on Ethereum are encrypted and pseudo-anonymous. The open and accessible nature of the network enables detailed data analysis, and anybody may examine and modify the source code for the Ethereum and DeFi protocols.
- **Permissionless:**DeFi, built on Ethereum, differs from conventional finance because it makes financial services available to anybody with a crypto wallet and Internet access.
- **Self-Custody**: DeFi market participants can communicate with permissionless financial applications and protocols using Web3 wallets such as MetaMask while preserving complete control over their assets and data.

3.3.4 Use Cases and Activities of Decentralised Finance

Trading: Trade Finance, commonly referred to as TradFi, was a centralised organisation in charge of organising and carrying out transactions between parties using an order book. On-chain currency trading has been possible since the advent of smart contracts, leading to the development of DeFi. Several intermediary trade administration functions, such as orderbook verification, online payment gateways, and stockbrokers, need to be updated by blockchain technology (i.e., users manage their assets while the blockchain manages to perform correct transactions). Derivatives, margin, and token swaps are traded over a linked network of exchanges, liquidity pools, and marketplaces in the DeFi space. The benefits of decentralised cryptocurrency exchanges include lower costs, faster transactions, and complete asset control. Unregulated Bitcoin exchanges are made possible by liquidity pools. Users have the option to conduct commerce without disclosing their names. LPs produce money in two currencies, enabling traders to trade pooled tokens. Automated market makers, or LP price trades, use a constant function market maker-like pricing function (CFMM). CFMM trades must change reserves to maintain the product.

Yield Farming: Money is routinely transferred from one interest-bearing pool to another throughout the "yield farming" process to deposit assets in the pools with the most significant annual percentage rates (APR) the following week. Money can then be invested in riskier tokens that might produce more significant returns (Cousaert, Xu, and Matsui, 2022). For instance, a farmer borrows $100,000 via compound financing. They will then get 100,000 cUSDT at a predetermined interest rate. A farmer with a loan in USDT can move to ETH the following week to lock in the higher rate when the interest rate on ETH increases (which may occur numerous times per week or month). These funds total 100,000 USD and cannot be used while frozen. Gains can be improved by locking the returned currency (cUSDT for compound) in an exchange pool on Balancer, an automated portfolio manager, and the Ethereum trading platform (Rahman et al., 2021, pp. 10–14). Users frequently pay a fee to make transactions on a decentralised exchange easier. For instance, the fee for each trade on VVS Finance is 0.3%. A portion of the platform's transaction fees will go to those who add to the platform's liquidity pool. Yield farming is the name for this type of agriculture.

Platforms that reward users with tokens for participating in particular activities are motivated to increase the quantity of those activities, which benefits yield farmers by boosting their earnings and lowering transaction costs.

A user who deposits money earns credit and then repeats the process with assets from a load to enhance their position. In other words, the interest received will multiply the initial payment. The user can mine for more liquidity by repeatedly borrowing and lending, which boosts their profits. Leverage raises the risk attached to these funds, though, as nominal price and debt ratio changes considerably influence the user's capital and may lead to their liquidation (Qin et al., 2021, pp. 3–32).

DEXs: DEXs, which allow peer-to-peer transactions with asset control, are the most popular DeFi applications. DEXs eliminate the risk of price manipulation, hacking, and theft and provide token projects with comparable liquidity to centralised exchanges without listing fees.

Unlike centralised exchanges, DEXs operate as automated market makers (AMMs) and enable transactions via liquidity pools. This means that users of decentralised exchanges can trade practically any token, but users of centralised exchanges are restricted to the tokens supported by the exchange.

On the DeFi market, popular DEXs include AirSwap, Liquality, Mesa, Oasis, and Uniswap. MetaMask Swaps, an aggregator of DeFi liquidity data, aids traders by offering the best price quote, optimal gas prices, and the lowest failure rates.

Tokenization: One of the pillars of decentralised finance, tokenization is built into the Ethereum network. In addition to powering the network, tokens provide access to new commercial opportunities. Tokens are digital assets generated, distributed, and managed on a blockchain. Tokens can be configured to perform many activities before the user accesses them, and they are designed to be transferable and secure. Tokens built on the Ethereum blockchain have become a popular way for people worldwide to access, exchange, and hold assets. They are used for everything from representing partial real estate ownership to incentivizing a specific programme.

Staking: Ethereum 2.0 will adopt Proof of Stake (PoS) as its consensus method, replacing Proof of Work (PoW). ETH holders can participate in staking and earn rewards in two different ways. As part of the staking mechanism, coins are deposited to prove a user's desire to participate in the block-building process. If users successfully create the next block, the network will compensate them with mining fees. In Ethereum, for instance, bets are made in increments of 32 ETH. If a user has staked a certain quantity of coins relative to the entire stack volume, then that person has a greater chance of being selected to attempt to generate the next block. Payments to miners are made every time they successfully construct a new block. Staking funds on Ethereum could be in danger if users' balances are reduced. The network levies a fine and deducts money from the user's account, possibly all of it. This situation will arise only if the user regularly fails to generate a good block.

Financial Savings: By altering the interest rate following market supply and demand, DeFi systems, such as lending pool protocols like Compound, provide interest-bearing accounts with better yields than conventional savings accounts. Another well-known method for customers to increase their return on investment is through savings accounts. A person can open a crypto savings account to store cryptocurrency or fiat money. These accounts are significantly different from traditional savings accounts in that they are not FDIC-insured. Instead, DeFi systems, such lending pool protocols like Compound, offer interest-bearing accounts with higher rates. Another well-known method for customers to increase their return on investment is through savings accounts. A person can open a crypto savings account to store cryptocurrency or fiat money. There is a significant difference in that these accounts are not FDIC-insured. Like most financial products, more significant gains come with greater risks. These accounts generally offer 8% interest, the industry standard for stock market returns (e.g., BlockFi, a cryptocurrency exchange and wallet platform). The ability to save money with cryptocurrency has dramatically improved. Argent, Dharma, and PoolTogether are three of the most well-known savings applications; they are all no-loss savings games where players get their initial investment back whether they win or lose.

Borrowing and lending: One of the most popular DeFi applications is P2P lending and borrowing protocols. Multiple DeFi systems, including PoolTogether, Argent, and Dharma, can be connected to and supported by Compound, an algorithmic interest rate protocol. Its Ethereum-based interest rate platforms enable borrowers to get interested in their cryptocurrency holdings and lenders to lend out their cryptocurrency assets. The compound smart contract successfully links borrowers and lenders and determines the interest rate based on the borrowed-to-supplied-assets ratio. The success of Compound demonstrates the DeFi space's enormous potential, and as more businesses implement the protocol, they will be able to generate interest from unused crypto assets.

In conventional finance, banks lend money based on the amount of money other account holders deposit into their savings accounts. DeFi lending follows a similar concept, except that fund offerings are deliberate and clients give liquidity to the lending market. Loan liquidity providers are reimbursed with interest on the funds as compensation. In contrast, borrowers have access to a broad market. The proportion of their available collateral defines a user's borrowing limit. If the borrowing limit is 75% of the value of the collateral, then the borrowing limit for $100 in collateral is $75.

4. Digital Currency: Pioneer of Cashless Ecosystem

Many believe the advent of the Internet will make physical money obsolete and shift all financial transactions online. Digital currency has become more popular recently, even though physical money used to be widely accepted. Checks and bonds have long been suggested as alternatives to cash. However, it was only possible to do away with paper money once mobile financial services and cryptocurrencies emerged.

4.1 Value of Money: Cash to Cashless

Gold and silver were used as currency because of their decorative and practical qualities. Commodities like gold and silver were also hard to come by. In the past, bank or central bank-issued paper money had value because it could be traded for precious metals. It has value because governments insist on being paid in fiat currency. Bank deposits may be valuable since they may be exchanged for "real" currency.

So, what makes virtual currency worthwhile?

Professor Eswar Prasad indicated in his 2021 book *The Future of Money* that Bitcoin and other cryptocurrencies cannot be swapped for traditional fiat currency. Bitcoin's finite quantity of 21 million coins is viewed by its proponents as the digital counterpart of gold. However, unlike gold, Bitcoin has no practical applications outside of its monetary worth.

David Gerard, in his book "Attack of the 50-Foot Blockchain" (2017), and Gary Silverman (2021), in their writings, stated the possibility that digital currency could be used by criminals to avoid detection, as well as to circumvent anti-money-laundering and know-your-customer measures, which raises serious concerns about the future of the industry. Again, alternatively, JP Koning and Kelly (2021) highlighted that cryptocurrencies had been criticised for their perceived similarities to decentralised Ponzi schemes or chain letters, which are said to be propelled by cult-like devotees and opportunists.

Stable coins, such as Tether or USD Coin, have value because they are exchangeable for dollars or a basket of currencies on a one-to-one basis.

This makes most stablecoins comparable to fiat currency. Tether has stored receivables and commercial paper instead of cash and near-cash assets.

There are two methods by which the government can regulate digital currency. One is that they mandate minimum capital and liquidity levels for cryptocurrency exchanges (treasury.gov, n.d.). "The main takeaway from creating private money in the 19th century is that stablecoins must be subject to the same regulations as banks" (Gorton and Zhang 2021, pp. 1–53). In the future, central banks might issue their own digital currency, known as "central bank digital currency" or CBDC, and establish a monopoly similar to what they did with real coins and paper currency.

4.2 Concept of Digital Currency

Digital money is a commodity that only exists in digital or electronic form. It is sometimes called cyber-cash, electronic currency, or electronic money. Internet-connected computers and digital wallets are used to conduct transactions. Unlike traditional physical currency, digital currency can be utilised for international transactions in online communities such as social networks, gaming websites, and gambling portals.

Decentralised and centralised digital currencies exist. The central bank and regulatory entities issue tangible fiat currency. Bitcoin and Ethereum are decentralised digital currencies.

Value can be communicated through digital currency. A paradigm shift is necessary to use digital currencies, linking them to buying and selling products and services. However, the principle is strengthened by digital currency. For instance, a gaming network token could potentially grant a player additional lives or more potent superpowers. This is a value exchange as opposed to a sale or purchase.

4.3 Types of Digital Currencies

The phrase "digital money" is sufficiently all-encompassing to incorporate the numerous kinds of virtual currencies now circulating. In general, there are three different forms of money, which are as follows:

Digital currency issued by central banks: Central banks issue and regulate CBDCs, a digital currency. Unlike conventional cash, which can be stored physically or digitally, cryptocurrencies are only available in digital form. The phrase "digital money" is sufficiently all-encompassing to incorporate the numerous kinds of virtual currencies now circulating. In general, there are three different forms of money, which are as follows: CBDCs facilitate the execution of monetary policy by eliminating intermediaries and establishing a direct link between the federal government and the ordinary citizen. The process no longer requires the participation of banks and financial entities responsible for the distribution of national currency.

CBDCs can be used and implemented in two ways. Similar to fiat currencies, retail CBDCs can be used for everyday purchases. Wholesale CBDCs are utilised by banks and financial organisations in limited applications.

Cryptocurrency: Cryptocurrencies employ cryptography to safeguard and verify transactions on a network. The cryptographic sheath surrounding digital currency increases the security and tamper-resistance of transactions. In addition, the creation of these currencies is regulated and controlled via cryptography. Cryptography is used to secure the cryptocurrencies Bitcoin and Ethereum. Depending on the region, cryptocurrencies may or may not be subject to regulation. The entire market capitalization and value of the cryptocurrency markets have dramatically expanded since 2017. This is because cryptocurrencies are becoming more accepted as an investment class. By July 2021, cryptocurrencies had a $2 trillion market valuation.

Virtual currency: Developers or a founding stakeholder organisation administer unregulated virtual currencies. Virtual currency can also be managed algorithmically using a well-established network protocol. Virtual currency can be demonstrated by a gaming network token controlled by the game's makers. Private issuers often produce virtual currencies and use them in specific online communities. It is crucial to ensure the safety of the infrastructure that supports virtual currencies, including the software and networks. There is no intrinsic value underlying virtual money. The sentiment of traders primarily determines a virtual currency's value. Due to its unregulated nature, a virtual currency is susceptible to significant price volatility.

4.4 Boons of Digital Currencies

Transaction at a Glance: Transactions using digital currencies are lightning-fast since all users are connected to the same network and no middlemen are involved. Transactions using digital currencies are typically rapid and inexpensive since they bypass the need for a third party to process the payment. This is superior to conventional payment systems involving banks or clearinghouses. Electronic transactions based on digital currency can facilitate record-keeping and transparency in business operations.

Monetary and Fiscal Policy Implementation: Digital currencies must include requirements for physical currencies, such as creating existing production facilities. New money is introduced into the economy by the Federal Reserve through the many intermediaries that are already present in the monetary system. These intermediaries include banks and other financial organisations. CBDCs can facilitate the circumvention of this system and the direct payment of citizens to government agencies. Additionally, they facilitate production and distribution by doing away with the prerequisites of creating money notes and transporting them between places.

Lower Cost Transaction: In the DeFi domain, numerous transactions can be conducted directly between users, eliminating the need for fee-charging intermediaries such as banks. This reduces costs and speeds up transactions. In addition, blockchain technology provides transparency because all transactions are recorded on a public ledger. Users have complete control over their assets and data, granting them greater privacy and security than conventional financial systems.

Zero Inflation: Bitcoin and other cryptocurrencies provide a decentralised alternative to conventional currencies and combat inflation with limited tokens. As currency demand increases and supply remains constant, the currency's value is anticipated to rise, resulting in deflationary pressure. This approach protects the currency's purchasing power over time, making it susceptible to market instability and demand fluctuations.

Decentralised and autonomous: Unlike conventional currencies, digital currencies are decentralised, implying that a single entity does not manage them. People have more control over their finances due to the lack of bank and government dominance. Like conventional wallets, digital currency wallets give users direct access to their holdings. This contrasts with the challenges encountered when moving fiat money between banks The security provided by digital currencies against fraud is one of their benefits. Businesses frequently keep personal and financial information while utilising traditional internet protocols for transactions, which makes them vulnerable to fraud if compromised. Users are not forced to divulge personal information because digital currencies enable anonymous and direct transactions.

Better accessibility with confidentiality: Accessibility is one of the digital currency's most prominent characteristics. Anyone with internet access can mine their favourite cryptocurrency or obtain dollars online. This contrasts with conventional institutions, which usually impose access restrictions like credit checks or minimum investments. The coin's transaction history is recorded and maintained, not that of the spender.

4.5 Risk measures associated with Digital Currencies:

Price Volatility: Since governments or central banks do not back digital currencies due to their decentralised nature, they tend to be unpredictable and subject to fast value fluctuations, both gaining and losing value swiftly. Digital currency traders can make a lot of money in a single minute. Because cryptocurrencies are decentralised, new digital currencies with low capitalizations and unreliable valuations have been created.

Other digital currencies have experienced a comparable early-stage price pattern. The price of Second Life's Linden dollars, which are utilised in online games, fluctuated initially.

Security Concern: Since digital currencies are computerised, they can be hacked. Online wallets might lose digital cash due to hacker attacks or have their protocols changed, making them useless. Criminals may take advantage of this readily available cash. Money laundering and the funding of unlawful activities are issues with digital currency.

Issues with storage and infrastructure: Even if traditional wallets are not required to use digital currencies, some processing and storage criteria must be met. For instance, a connection to the Internet, mobile phones, and the services associated with their provisioning are all required. It is necessary to use trustworthy online wallets to store digital currency.

Inadequacies in regulation: In the traditional financial industry, extensive regulation maintains stability and prevents unwanted behaviour. Digital currencies, on the other hand, operate primarily outside of such control, offering greater freedom but also rendering them susceptible to undesirable practises.

Unpredictable future: Traditional payment mechanisms have demonstrated their dependability throughout time and have a promising future. Digital currency, on the other hand, is still in its early phases of usage and development. There is no assurance that these assets will continue to be useful or preserve their value, which may result in an unknown outcome.

5. Interconnection between blockchain, AI, and IoT in securing digital currency

The goal of combining technologies such as blockchain, the Internet of Things (IoT), and artificial intelligence (AI) is to make the most of the potential offered by each technology. The Internet of Things (IoT) can gather and supply data, blockchain can provide the infrastructure and set the rules of engagement, and AI can optimise processes and rules; these things can act as connectors between these technologies (Salah et al., 2019; Zheng et al., 2020, pp. 95–106). They are shown in Figure 6. Data standards, privacy, security, and sustainability will all improve due to the convergence of these technologies, which will be helpful for the management of data and identities and automate business processes.

As the IoT becomes industrialised, many sensory devices produce enormous quantities of sensory data (Chung et al. 2017, pp. 295–96). As a result, the importance of big data analytics for Internet of Things applications is growing. Several academics advocate artificial intelligence (AI) technology for the Internet of Things to overcome these challenges. In big data analytics, AI fuels real-time data processing. The Internet of Things and artificial intelligence are combined to facilitate the collection and analysis of as much data as possible. Centralised hierarchies, security and privacy laws, economic constraints, and a scarcity of training data make designing and developing successful extensive data analytics systems using artificial intelligence difficult. It is depicted in Figure 8.

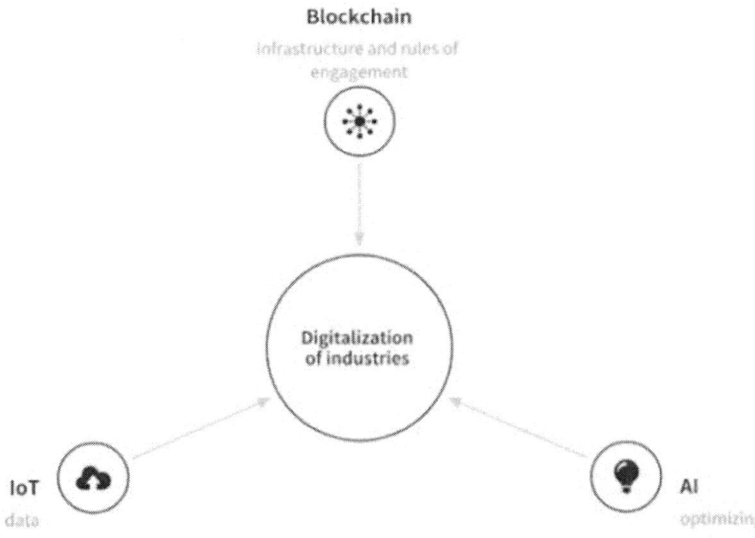

Figure 8: Convergence of technologies in digital transformation

In recent years, combining AI and blockchain has been instrumental in tackling numerous difficulties. By analysing vast amounts of data, blockchain and AI technologies have enabled organisations to overcome database difficulties. AI programmes are among the most in-demand technologies of the Fourth Industrial Revolution, as they can execute error-prone activities and reduce the need for manual labour (Sandner, Gross, and Richter 2020). Blockchain technology's decentralised nature facilitates the secure transmission of data and resources between many nodes in an IoT network, reducing the need for centralised control. It tackles typical AI-related challenges by combining blockchain technology into a decentralised AI system. Blockchain enables the secure transmission of encrypted signature data without the need for third parties (Ai and March 2018). In Internet of Things applications, devices can make decisions autonomously (Corea 2019, pp. 69–76).

The vital breakthrough areas that the triad can revolutionise shortly are depicted in Figure 7.

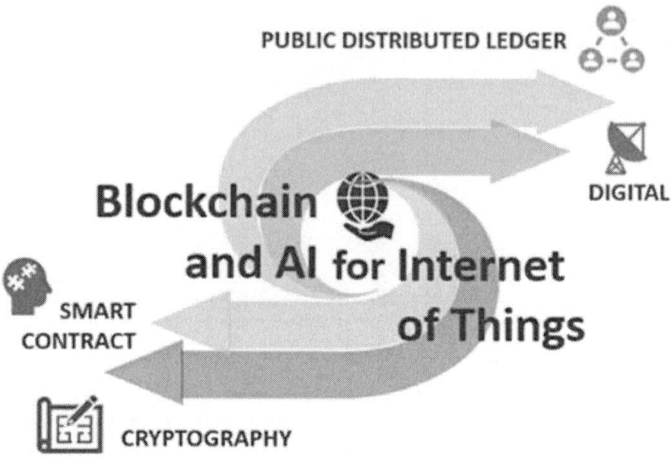

Figure 9: Blockchain, AI, and IoT in a Nutshell (Ankit Som and Parthajit Kayal, the authors of AI, Blockchain, and IoT, compiled the source material.)

Data management

Data standardisation: IoT devices generate enormous amounts of unstandardized data due to older data gathering and storage systems. To facilitate the standardisation of data, a single, user-accessible digital platform for IoT data may be created using blockchain technology. Blockchain systems use hash functions to maintain data uniformly, making them incredibly standardised. Hash functions minimise the quantity of data to be encrypted by transforming it into a preset string.

Data privacy and security: The encryption that underpins blockchain technologies allows for a great deal of confidentiality to be maintained. The blockchain technology behind cryptocurrencies like Bitcoin and Ethereum allows for anonymous transactions. However, enabling fully anonymous transactions is also possible, as with cryptocurrencies like Monero and Zcash. This is possible because blockchain systems use a private or public fundamental architecture that lets data be encrypted from end to end while at rest and in transit. Internet of Things devices and equipment typically capture and store personally identifiable information. Data storage privacy and security must be prioritised. IoT data is typically transmitted straight from the device to the corresponding database.

However, the lack of encryption means this information could be more secure.

However, there is a conflict between having a lot of privacy and being able to stop illegal things from happening. When transactions are anonymous, it is impossible to determine who sent them or where they live. Crimes like money laundering and terrorism financing benefit from this anonymity. Artificial intelligence (AI) can be helpful here since it can detect illegal activity, increasing security. (Yin et al. 2019, pp. 37–73) propose that artificial intelligence (AI) that uses data analytics might help mitigate the potential for criminal activity on the blockchain, which arises because of the platform's inherent privacy features. It is important to remember that the abundance of IoT data produced is a boon to AI technology since AI algorithms learn by observation.

Scalability: The large volume of data created by the Internet of Things may now be processed with additional resources. Integrating blockchain and AI has the potential to improve scalability substantially. Liu, Yuan, and Yu (2021, pp. 12690–710) suggest a framework for enhancing the performance of blockchain-enabled IoT devices. Machine learning could provide the basis for this system. According to the study's authors, the most efficient technique for enhancing performance is a "DRL-based system to dynamically select or adjust the block producers, consensus algorithm, block size, and block interval."

Identity Management

Blockchain-based authentication: The identity of Internet of Things (IoT) devices can be controlled using blockchain technology, which can be employed for authentication and consequently increase network participant confidence. Although individuals and organisations are typically the primary goals of identity management initiatives, IoT sensors and machines are frequently included. Individual identification documents, such as driver's licences, passports, and corporate registry entries, will be used to construct digital identities on the blockchain. The processing of transactions can be made more efficient and cost-effective by using these identifiers. According to projections made by IoT Analytics, over 20 billion devices will be connected to the Internet by 2025. Before joining this new payment network, individuals, businesses, and even robots must validate their digital identities. Blockchain technology is exceptionally well suited for establishing and managing digital IDs safely and efficiently. The identity system based on blockchain technology must adhere to data protection rules like traditional centralised systems. Blockchain technology's built-in access systems and encryption procedures make it better than non-blockchain-based systems for three primary uses: (i) securing data by design; (ii) managing data ownership; and (iii) promoting data monetization. Blockchain protects identity with immutable, unforgeable records.

Automatization

Smart contract automation: Even though smart contracts have great promise for increasing productivity in many fields, they have yet to acquire widespread acceptance. Because traditional smart contracts necessitate cryptocurrency, most businesses hesitate to adopt crypto assets due to regulatory and economic constraints. The volatility in value is the primary disadvantage of crypto assets. Due to the asset's high price volatility, the counterparty to a smart contract denominated in crypto assets faces severe exchange rate risk. Stablecoins are, in other words, coins that maintain a reasonably constant value throughout time, but some industrial firms may only adopt them. The first difficulty is that stablecoins still need to be supervised by a defined framework, causing cautious enterprises to avoid them. Second, traditional monetary units like the Euro and the US Dollar are used in most IT and accounting systems of businesses rather than crypto assets. As a result of the time and money it takes to switch from stable coins to fiat currency, this practise is inconvenient for bookkeeping.

1. **The Blockchain Euro**: Smart contracts can only reach their full potential in one manner: blockchain-based fiat money is required for the smart contract to "flow through." IoT devices cannot provide services like pay-per-use billing, leasing, and factoring independently; only a digital Euro backed by the blockchain would make that possible. A blockchain-based digital currency permits unique business models like fully autonomous gadgets that utilise AI to make judgements and "economically thrive" using blockchain for financial transactions and profit centre logic at the device level. Blockchain-based digital currencies can make IoT device micropayments easy and cheap.
2. **Central Bank digital currencies:** New research from the Bank for International Settlements (Barotini and Holden 2019, pp. 1–15) indicates that more than seventy central banks examine the potential repercussions of CBDC. However, no government entity has yet issued such a currency. In any event, governments are

beginning to interact with digital currencies; for example, the European Central Bank (ECB) recently developed "EUROchain," a CBDC prototype built on the Corda DLT framework. A CBDC issued by the ECB would make central bank money usable in blockchain-based smart contracts (Sandner, Gross, and Richter 2020).

Commercial banks issue e-money, also recognised as commercial money, as the central bank issues central money. Central bank money is not at risk of loss if the central bank goes bankrupt, unlike commercial bank money. When economic conditions deteriorate and financial institutions like banks and e-money sites are at risk of failing, this distinction becomes crucial.

6. Application of blockchain, IoT, and AI in securing finance

Blockchain, IoT, and AI can be coupled in many ways. Innovative technology is driving the rapid evolution and disruption of the financial industry. When these technologies are combined, they will enhance business practises, products, and services while accelerating the expansion of digital transformation in industrial businesses. To benefit from increased productivity, executives must address these technologies. The Internet of Things, artificial intelligence, and blockchain technology will usher in a new digital era.

It is sometimes said that FinTech is the "wedding" of technological advancement and economic theory. When resources (such as money) are paired with technological advances (such as new knowledge), the result is a multiplier effect. When resources (such as money) are paired with technological advances (such as new knowledge), the result is a multiplier effect. According to Fuentes (2017, pp. 1–14), few significant developments distinguish contemporary FinTech. Big data, deep learning, technical standards, and artificial intelligence (AI) increase variety. Second, nonfinancial corporations invest in banking subsectors. FinTech has two classifications. Technology impacts traditional financial organisations in the first segment. Traditional financial institutions, such as Morgan Stanley, the Industrial and Commercial Bank of China, and Goldman Sachs, use big data and maybe other new technologies to improve and adapt their existing services. FinTech has had an impact on the conventional banking system. Following the 2008 Financial Crisis, general economic laws and the financial industry's growth altered the scene (Brem, Bilgram, and Marchuk 2019, pp. 348–60). Mobile wallet solutions like WhatsApp Payment, Alipay, and Apple Pay are among the most innovative applications of FinTech. Then, there is a "smart contract" that includes firms like "Ant Xiaodai," "Huabai," and "Jingdong Baitiao." Blockchain, which has already been mentioned, is the final concept. The foundation of these three critical FinTech sectors is quick communication, current data, reliable credit, and fast alarms. Blockchain technology has captured the banking industry's attention because it can alter the institutional landscape and allow people to build trust quickly (Banafa 2020, pp. 1–154).

Fiat currency based on the blockchain and governed by the contract is required to fully realise the potential of smart contracts. Smart contracts utilising Euro values can only be executed via a blockchain. Fiat currency automatically enables machines, cars, and sensors to offer pay-per-use, leasing, and factoring services. A digital Euro based on blockchain technology could result in autonomous robots that employ artificial intelligence to make choices and execute financial transactions and corporate operations using blockchain technology. The Euro can be issued by financial institutions, e-money providers, unregulated firms, and central banks. Several Chinese cities, including McDonald's and Starbucks, are currently testing the Chinese CBDC project DC/EP. Using a central bank-backed Euro in European industrial smart contracts would demand a blockchain-based CBDC, although the European Central Bank has yet to declare the issuing of CBDCs (Sandner, Gross, and Richter 2020).

Companies are investigating the possibility of combining AI and blockchain to develop more valuable goods due to these two technologies' growing popularity and numerous benefits. The distributed ledger technology known as blockchain is a user-friendly, transparent, energy-efficient, immutable digital ledger administered by artificially intelligent agents to ensure the highest possible level of security. State Street is developing an artificial intelligence system to verify investment data and an immutable blockchain with cryptographically proven integrity. According to Trent McConaghy (founder and CTO of BigchainDB), many other businesses, including significant technology firms like IBM, invest in related research.

On December 30, 2015, NASDAQ (National Association of Securities Dealers Automated Quotations) stated that Chain.com had completed and documented a private transaction utilising their Linq blockchain technology. This news came as a surprise to the company. This transaction was the first to use blockchain technology, which provided the issuer with a digital ownership record, shortened the time needed for settlement, and avoided the demand for paper stock certificates. Open Assets Protocol is a coloured currency invention produced utilising the blockchain. It was integrated into the Nasdaq Private Market platform when the company launched a blockchain-enabled solution to strengthen and expand the equities management capabilities of the platform.

Using the R3 Corda distributed ledger technology, financial contracts may be tracked, controlled, and coordinated between regulated financial institutions. The time and effort required to reconcile data manually across numerous databases will be reduced. It would also boost financial code sharing, decreasing costs for everyone.

What if your bank account controlled your thermostat instead of your home's weather data? Imagine if your thermostat could be set to lower temperatures automatically whenever your savings fell below a certain level. What if there is already such a choice? Indeed, it does. The Interact IoT platform, created by a London-based IT firm Digital lets financial institutions link consumer accounts to consumers' smart gadgets. Devices of this type can then react to fluctuations in user account balances.

The evolution of the Internet of Things depends on artificial intelligence (AI) and its forerunner, blockchain. Through collaboration with those who process claims, data scientists provide their machines with new learnings that improve their capacity to identify fraudulent activity. Watson is used by H&R Block to assist its tax preparers in understanding and applying the tax code and by KPMG to uncover data abnormalities in its audit business. AI can transform every industry it enters because it recognises patterns, mistakes, anomalies, and other features in all these contexts.

Therefore, businesses must jump on board with these new technologies and imagine the future. Instead of seeing new technologies as a threat, finance departments can see them as an opportunity to play a more central role in moving the company ahead and implementing innovative new approaches to doing business.

7. Unlocking the Future with Digital Finance

It is true that we are living in a digital age and that banks and other financial institutions must now prioritise digitization to continue to thrive. The Internet of Things is essential for data and other vital information.

Blockchain technology validates and tracks data by building infrastructure to speed up data transfer processing, and AI optimises the entire process.

7.1 Digital finance and its impact on financial sustainability

Emerging economies, particularly those in low-income and remote locations and small and medium-sized firms, are usually disregarded by conventional financial institutions and have a long-standing problem with financial crises. This issue is particularly real for economies in the development process located in regions with poor economic conditions and remote locations. According to a survey undertaken by the World Bank, millions of formal and informal small and medium-sized enterprises in developing economies have a loan demand of around $8.1 trillion. This loan demand figure is equivalent to 40% of the GDP. The loan demand accounts for around one-third of all adults on the face of the Earth. These findings help to explain why it has received such a significant amount of attention ever since the United Nations (UN) initially adopted the phrase "financial inclusion" in 2015. A stable financial climate is essential for developing financial services over time, making them more inexpensive, accessible, and of higher quality overall. The provision of access to financial services is one of the benefits of financial inclusion for low-income populations.

The rise of digital finance has played a crucial role in broadening access to banking services and improving the lives of those in poverty, even in economically disadvantaged locations. This positive impact has been made feasible by concurrently developing other cutting-edge technologics, like big data, cloud computing, AI, IoT, and blockchain, which frequently complement one another. As of 2019, according to the Global Association for Mobile Communications Systems (GSMA), around one billion individuals worldwide have access to various financial services. In addition, the International Monetary Fund (IMF) believes that mobile payments facilitated by inclusive digital finance contribute around 2% to global GDP growth.

Financial sustainability is achieved when the benefits of financial services surpass their drawbacks; increasing the effectiveness of financial operations helps achieve this goal while simultaneously reducing expenses and boosting benefits. Improving one's financial effectiveness is essential to maintaining one's financial standing. The effects of financial technology can be seen on both a macro and micro scale. Studies have shown that using digital finance positively impacts conventional banking practises, the growth of the economy, and the distribution of money more fairly between urban and rural areas. Most studies investigating digital finance's influence on small and medium-sized enterprises (MSMEs), local entrepreneurship, domestic financial requirements, and household spending have concentrated on micro-level research. This focus on micro-level research shows that it may substantially influence satisfying the wants of "long-tail" consumers.

7.1.1 Accelerating Financial Efficiency: Digital Finance's Key Drivers

Transforming Financial Accessibility through Online Coverage: The market size for financial services is proportional to the number of consumers. To use most conventional forms of finance, you need to have an account with that financial institution, and opening one for the first time requires a trip to the bank's physical location. This requirement implies that physical banks will have to keep opening additional locations to meet the needs of their customers. While reaching out to consumers at the end of the economic tail can result in positive cash flow for traditional banks, the substantial expense of maintaining physical locations could result in a loss. Digital finance reduces the cost of financial services and eliminates geographical limitations through the rapid transmission of data across great distances. This capability allows financial institutions to serve a wider range of customers whose benefits outweigh their costs, therefore restoring the financial viability of the institutions (Luo, Luo, and Lv, 2022).

The level of digital financial adoption: Most users must conduct financial transactions in person at traditional banking institutions. By leveraging digital finance, users can conveniently access a diverse range of online financial services without the need to visit a physical branch, effectively bypassing geographical restrictions. This is facilitated by the rapid transmission of information across various locations made possible by digital finance. Due to its enormous information gathering and processing capabilities and low marginal costs, financial institutions can use digital finance to combine financial services into everyday circumstances. It can also facilitate providing financial services uniquely suited to the client's requirements (Andersen and Benzoni 2006, pp. 1–4).

7.1.2 Resistances in the Digital Finance and Financial Service Sectors towards ensuring financial inclusion:

Awareness of digital financial products and services has become as important as moving to a digital system. Increasing one's knowledge of DFS increases one's likelihood of adopting a digital financial services system, contributing to the aim of inclusive growth. Knowledge of basic concepts related to digital financial services, such as credit cards and debit cards, interest rates, inflation, the time value of money, risk perceptions, online payment systems, and SIPs, is included in this category. Due to the complexity of financial institutions' efforts to streamline their product offerings, there have been very few global surveys of students' knowledge of financial services (Rai and Sharma 2014, pp. 600–607). While there may be a favourable association between user-friendly digital banking and financial inclusion, the relationship may need to be strengthened and made advantageous for low-income and disadvantaged digital finance users. Fear of new technology, high expenses, financial awareness, and other issues may prevent marginalised people from adopting digital financial services.

Additionally, service providers may discriminate by not actively marketing their digital finance services to low-income and impoverished clients, resulting in diminished financial inclusion. The monetary payoff for digital finance providers is more significant with high- and middle-income customers, resulting in a less aggressive marketing strategy for low-income clients. Based on their evolving internal risk assessment, some providers may limit delivery of digital money services to high-risk locations requiring more infrastructure. In addition, online banking services may be more suited to those with superior cognitive talents. If digital finance providers assess that reaching low-income communities would yield a negative return on investment, they may decide to curtail their outreach efforts. This is especially true when individuals require additional information and expertise to use and understand digital money effectively. (Ozili 2018, pp. 329–40).

Financial services firms should consider investing in employee training to keep up with rapid technological change. In addition, as part of the initial rollout, businesses may need to bring in outside experts to guarantee that their implementations align with all audit, internal, and regulatory requirements.

Although companies invest heavily in making their systems as safe as possible, introducing new technology always carries some danger. Because of this, financial institutions need to exercise increased prudence and continue doing risk assessments of the blockchain-based smart contracts and Internet of Things (IoT) devices they have installed. When conducting business financially online, one of the most serious concerns that needs to be addressed is how to keep personal information secure. Researchers are always looking for new ways to ensure technology can satisfy regulatory requirements, and privacy is a growing part of the financial ecosystem.

7.2 Digital Finance: Shaping the Future of Trade Finance

The nature of trade has remained unchanged for generations. This is because trade fundamentally revolves around the transaction between a buyer and a seller. However, the issue is more complex in the modern world, both commercially and monetarily.

The trade sector is a significant driver of global economic growth, which accelerates as global demand expands. The demand for financing solutions, particularly from local providers, has expanded due to the expansion of the global trade industry, which has led to a noticeable increase in the participation of rising local markets in global commerce. Since the flow of products and services across borders is only possible with trade finance, local financial sectors must develop swift and flexible means of financing their trade and adapting to the sudden needs of the market. As the pandemic's effects ripple through economies, a $6.5 trillion trade finance deficit has become a permanent fixture. Banks, and especially trade finance departments, are trying to find ways to reduce the risks they take on through the use of innovative solutions that improve the efficiency of trade finance departments and allow them to leverage their businesses better in order to better meet customer demands, mitigate the adverse effects of new regulations, and, ultimately, keep trade finance funding from being cut.

Data transmission and payment processing durations will decrease when the bank's internal procedures are digitised and improved. These advancements will make it possible to meet the market's financial demands considerably more rapidly while reducing the potential for fraud through more secure transaction procedures.

8. Conclusion

While it is now commonplace to conduct financial dealings online or through a mobile device, future consumers should proceed with caution because of the inherent risks associated with these methods. Based on the premise that currency serves as a record of economic activity, two paths present themselves for developing the electronic currency: first, guaranteeing the security and reliability of the financial system, and second, the irrevocability of monetary transactions. Methods for accomplishing this can be broken down into two camps: those favouring a centralised intermediary and those favouring a decentralised governance structure. Furthermore, regulations need to be established for who can access this information and under what conditions, with proper safeguards in place to ensure confidentiality. That is to say, we must ensure the payment mechanism is verified and private. When integrated, blockchain, IoT, and AI improve transaction data security, transparency, immutability, and privacy, making financial automation easier. By obtaining a higher degree of uniformity, we can raise the standard of data management and guarantee data privacy and security.

References

Ai, Nebula, and Team March. 2018. "Nebula AI (NBAI) — Decentralised AI Blockchain Whitepaper." Nebula AI Inc., NEBULA, https://neironix.io/documents/whitepaper/4082/NBAI_whitepaper_EN.pdf

Andersen, Torben G., and Luca Benzoni 2006. "Federal Reserve Bank of Chicago." October 63 (210): 1–4. http://www.chicagofed.org/webpages/publications/chicago_fed_letter/2006/april_225a.cfm.

Anderson, Billie. 2019. "Using Bayesian Networks to Perform Reject Inference." *Expert Systems with Applications,* 137: 349–56, https://doi.org/10.1016/j.eswa.2019.07.011.

Article, Title, n.d. "Manuscript Details Manuscript Number JEB_2017_104_R2 Title FinTech and RegTech: Impact on Regulators and Banks."

Atlam, Hany F., and Gary B. Wills. 2020: *IoT Security, Privacy, Safety, and EthicsInternet of Things* Springer International Publishing https://doi.org/10.1007/978-3-030-18732-3_8.

Atlam, Hany F., Ahmed Alenezi, and Madini O. Alassa. 2020. "Security, Cybercrime, and Digital Forensics for IoT." in Principles of Internet of Things (IoT) Ecosystem: Insight Paradigm. https://doi.org/10.1007/978-3-030-33596-0.

Atlam, Hany F., Robert J. Walters, and Gary B. Wills 2018. "Internet of Things: State-of-the-Art, Challenges, Applications, and Open Issues," 9 (3): 928–38.

Banafa, Ahmed. 2020. "Blockchain Technology and Applications." *Blockchain Technology and Applications*, 1–154. https://doi.org/10.4018/978-1-7998-3473-1.ch085.

Barbu, Andreea, Gheorghe Militaru, and SavuIonut 2020, "Factors Affecting the Use of Smartwatches." *FAIMA Business & Management Journal* 8 (1): 44–57. https://eds.p.ebscohost.com/eds/detail/detail?vid=0&sid=766296cd-6104-4d46-a482-520fd7fd8018%40redis&bdata=Jmxhbmc9a HUmc2l0ZT1lZHMtbGl2ZQ%3D%3D#AN=143298887&db=buh.

Barotini, Christian, and Henry Holden "Proceeding with Caution: A Survey on Central Bank Digital Currency." *Bank for International Settlements* 101 (101): 1–15

Brem, Alexander, Volker Bilgram, and Anna Marchuk. 2019. "How Crowdfunding Platforms Change the Nature of User Innovation: From Problem Solving to Entrepreneurship." *Technological Forecasting and Social Change* 144 (October 2017): 348–60. https://doi.org/10.1016/j.techfore.2017.11.020.

Buterin, Vitalik. 2009. "A Next Generation Smart Contract and Decentralised Application Platform," no. January: 1–36.

Chung, Che Min, Cai Cing Chen, Wei Ping Shih, Ting En Lin, Rui Jun Yeh, and Iru Wang. 2017. "Automated Machine Learning for the Internet of ThingsLin, Rui Jun Yeh, and Iru Wang. 2017. "Automated Machine Learning for the Internet of Things" *2017 IEEE International Conference on Consumer Electronics, Taiwan, ICCE-TW 2017*, 295–96
https://doi.org/10.1109/ICCE-China.2017.7991112.

Corea, Francesco. 2019. "AI and Blockchain." *Studies in Big Data* 50: 69–76. https://doi.org/10.1007/978-3-030-04468-8_11.

Cousaert, Simon, Jiahua Xu, and Toshiko Matsui 2022. "SoK: Yield Aggregators in DeFi." *IEEE International Conference on Blockchain and Cryptocurrency, ICBC 2022*
https://doi.org/10.1109/ICBC54727.2022.9805523.

Fuentes, Marine Marta Martos. 2017. "No 主観的健康感を中心とした在宅高齢者における健康関連指標に関する共分散構造分析Title," 1–14.

Goel, Nikita. 2018. "Security Issues in Blockchain and Crypto Currency Security Issues in Blockchain and Crypto Currency," no. April.

Gorton, Gary B., and Jeffrey Zhang 2021. "Taming Wildcat Stablecoins." *SSRN Electronic Journal*, 1–53. https://doi.org/10.2139/ssrn.3888752.

Karafiloski, Elena, and AnastasMishev. 2017. "Blockchain Solutions for Big Data Challenges: A Literature Review." *17th IEEE International Conference on Smart Technologies, EUROCON 2017: Conference Proceedings*, no. July: 763–68 https://doi.org/10.1109/EUROCON.2017.8011213.

Liu, Tao, Yi Yuan, and Zhongyang Yu. 2021. "The Service Architecture of the Internet of Things Terminal Connection Based on Blockchain Technology." *Journal of Supercomputing* 77 (11): 12690–710. https://doi.org/10.1007/s11227-021-03774-9.

Luo, Dan, Man Luo, and Jiamin Lv. 2022 "Can Digital Finance Contribute to the Promotion of Financial Sustainability? A Financial Efficiency Perspective." *Sustainability (Switzerland)* 14 (7) https://doi.org/10.3390/su14073979.

Ozili, Peterson K. 2018. "Impact of Digital Finance on Financial Inclusion and Stability." *Borsa Istanbul Review* 18 (4): 329–40. https://doi.org/10.1016/j.bir.2017.12.003.

Perera, Charith, and C. H. I. Harold Liu 2015. "A Survey on the Internet of Things From an Industrial Market Perspective." 2.

Perera, Charith, Chi Harold Liu, and Srimal Jayawardena 2015. "The Emerging Internet of Things Marketplace From an Industrial Perspective: A Survey," 6750 (c): 1–13. https://doi.org/10.1109/TETC.2015.2390034.

"Perspective Blockchain Adoption in." n.d.

Pflaum, Alexander A., and Florian Michahelles 2018. "The IoT and Digital Transformation: Towards the Data-Driven Enterprise," no. March: 87–91.

Qin, Kaihua, Liyi Zhou, Benjamin Livshits, and Arthur Gervais 2021. "Attacking the DeFi Ecosystem with Flash Loans for Fun and Profit" *Lecture Notes in Computer Science (Including Subseries Lecture Notes in Artificial Intelligence and Lecture Notes in Bioinformatics)* 12674 LNCS: 3–32 https://doi.org/10.1007/978-3-662-64322-8_1.

Raheman, Ali, Anton Kolonin, Ben Goertzel, Gergely Hegykozi, and Ikram Ansari 2021. "Architecture of an Automated Crypto-Finance Agent." *2021 International Symposium on Knowledge, Ontology, and Theory, KNOTH 2021*, 10–14 https://doi.org/10.1109/KNOTH54462.2021.9686345.

Rai, Kamini, and Mamta Sharma. 2014. "A Study on Awareness about Digital Financial Services among Students," 600–607.

Reyna, Ana, Cristian Martn, Jaime Chen, Enrique Soler, and Manuel Daz 2018. "On Blockchain and Its Integration with IoT. Challenges and opportunities." *Future Generation Computer Systems* 88 (2018): 173–90. https://doi.org/10.1016/j.future.2018.05.046.

Salah, Khaled, M. Habib Ur Rehman, Nishara Nizamuddin, and Ala Al-Fuqaha 2019. "Blockchain for AI: Review and Open Research Challenges." *IEEE Access* 7 (c): 10127–49. https://doi.org/10.1109/ACCESS.2018.2890507.

Sandner, Philipp, Jonas Gross, and Robert Richter 2020. "Convergence of Blockchain, IoT, and AI." *Frontiers in Blockchain* 3 (September) https://doi.org/10.3389/fbloc.2020.522600.

Schär, Fabian. 2021. "Decentralised Finance: On Blockchain- and Smart Contract-Based Financial Markets." *Federal Reserve Bank of St. Louis Review* 103 (2): 153–74. https://doi.org/10.20955/r.103.153-74.

Som, Ankit, and ParthajitKayal. 2020. "AI, Blockchain, and IOT," 141–61.

Son, H., C. Hyun, D. Phan, and H. J. Hwang 2019. "Data Analytic Approach for Bankruptcy Prediction." *Expert Systems with Applications* 138: 112816. https://doi.org/10.1016/j.eswa.2019.07.033.

Sultan, Karim, Umar Ruhi, and Rubina Lakhani 2018. "Conceptualising Blockchains: Characteristics and Applications," 49–57.

Sundmaeker, Harald, and Antoine de Saint-Exupéry. 2010. *Vision and Challenges for Realising the Internet of Things*

Wood, Gavin. 2014. "Ethereum: A Secure Decentralised Generalised Transaction Ledger." *Ethereum Project Yellow Paper*, 1–32.

Yin, Hao Hua Sun, Klaus Langenheldt, Mikkel Harlev, Raghava Rao Mukkamala, and Ravi Vatrapu 2019. "Regulating Cryptocurrencies: A Supervised Machine Learning Approach to De-Anonymizing the Bitcoin Blockchain" *Journal of Management Information Systems* 36 (1): 37–73 https://doi.org/10.1080/07421222.2018.1550550.

Zheng, Peilin, Zibin Zheng, Jiajing Wu, and Hong-Ning Dai, 2020. "XBlock-ETH: Extracting and Exploring Blockchain Data From Ethereum." *IEEE Open Journal of the Computer Society* 1 (4): 95–106 https://doi.org/10.1109/ojcs.2020.2990458.

CHAPTER 6

FUTURISTIC SECURED TRANSACTIONS USING BLOCKCHAIN

B. NAGAJAYANTHI

VELLORE INSTITUTE OF TECHNOLOGY, CHENNAI, INDIA

Abstract

Advancements in businesses, academics, industries, healthcare, and government organisations have led to a stringent need to manage assets and data. Blockchain has gained momentum in the Industry 4.0 era due to its security and transparency. Blockchain has information in the form of a chronologically organised and cryptographically linked set of blocks. These are used to validate and record digital transactions without the need for trusted third-party authentication. The blockchain node serves as a decentralised shared digital ledger that archives all the current transactions and makes the data available to all connected devices. The digital data is in the form of blocks. These blocks are cryptographically linked to the previous blocks to form the blockchain. New information is added along with the previous blocks and compiled as a new block. Information could be of different types. Blockchain is preferred for transactions occurring between globally distributed computers. This technology is implemented in IoT, banking sectors, healthcare, supply chain, tracking assets, and so on., due to its data integrity. Blockchain secures transactions with its reliability, speed, and immutability. In this research, the potential implications of blockchain are reviewed. The challenges involved in the trending sectors and the thwarting measures of blockchain are demonstrated using use-cases. Originating from cryptocurrency, blockchain has leaped into trending sectors due to its security concerns and robustness. Most of our social media data is stored centrally and is at risk. Hence the need for decentralised, secured storage. Ethereum, a digital currency, uses blockchain for its security.

Keywords: data integrity; cryptography; blockchain; Ethereum

Introduction

For years, people have been managing data using ledgers. With the advancements in technology and the ability to track and synchronise data, digital information has moved from papers to computers. Technology has a booming impact on banking transactions. Banking is a major sector in the global economy. With an increased number of users, it is hard to maintain transaction data. With digitization, there is a stringent need for the banking sector to monitor transactions. In the digital era, with centralised Internet sharing, there are security breaches and privacy leakages (Chuan et al., 2022, 10). Currently, banking transactions are centralised and liable to tampering. This could be subject to counterfeiting and fraud. With digital innovations and financial transformations, data needs to be secured (Thulya, 2020, 7). With inclusions in automatic teller machines, electronic fund transfers, mobile banking, etc., blockchain is a game changer in the banking industry. Blockchain provides solutions to fraudulent transactions, identity theft, money laundering, and so on, with immutability, transparency, privacy, and security. Blockchain has been in existence since 2009 to keep data synchronised across multiple independent stakeholders. Eventually, the bitcoin cryptocurrency was created as the first decentralised electronic cash system. Blockchain consists of five entities: the block sequence, chain sequence, cryptographic linking, distributed replicated configuration, and the consensus mechanism. Each block has an identity, an ordered recursive chain sequence in which each block has a trace of all the previous recorded blocks, a cryptographic process in each block using a unique hash function to record the transactions, and a timestamp indicating the block creation. In blockchain, each block has the hash of the previous block. All the blocks are cryptographically linked. Hence, if the data is modified, the hash function changes, and subsequently, tampering is identified. In the distributed configuration, each node stores a portion of the blockchain, whereas the replicated configuration stores the entire copy of the blockchain. The consensus mechanism provides guidelines for the rules to be adhered to on the blockchain. Universal blockchain in banking estimates a turnover of around $4112.5 million by 2029. The secrecy and privacy of individual transactions are preserved without the intervention of a third party. Blockchain maintains a permanent digital ledger of previous transactions in the form of blocks. Blockchain progresses even if either of the nodes fail. Blockchain works on both centralised and decentralised systems due to its operational efficiency.

Blockchain Structure

While money is transferred using online banking, the transaction is updated in the records. This is simple but could be tampered with. Blockchain is integrated into many such financial transactions due to its triple combination, namely: cryptographic keys, a peer-to-peer network with a shared ledger, and a computing mechanism to store and record the transactions. Blockchain uses cryptographic keys as a digital signature to perform digital transactions. Servers and the cloud are replaced by a network of systems called nodes. Nodes store and share databases. The commercial and banking industries have technically adopted blockchain to a major extent. Blockchain is used to store data securely (Jayasudha et al. 2020). The steps involved while using blockchain during the bank transaction are as follows:

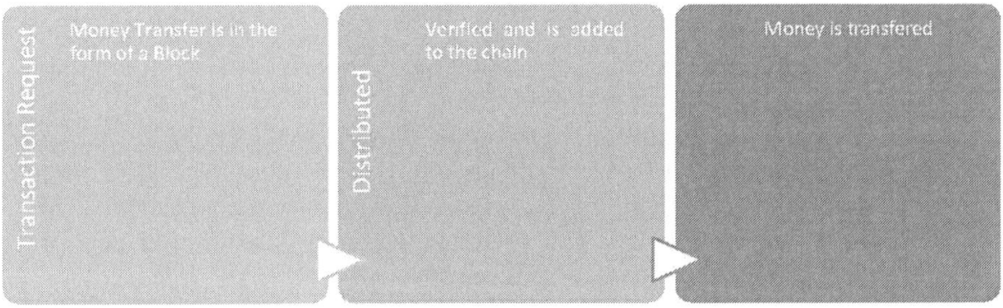

Figure 1. Blockchain Structure

- Information is represented as a block. When a transaction needs to be performed, a new transaction request enters the network. Information is encrypted and sent as a block. This is sent to all the distributed devices.
- **Validation**: Several transactions occur at the same time. These blocks are distributed and validated. Once confirmed, the block is added to the digital ledger.

Hash Algorithm: Functionally similar blocks are combined to form a blockchain. The hash code takes the data length as the input and generates an alphanumeric hash as the output, which serves as a digital fingerprint. Each block depends on the hash value of the previous block, as shown in

Figure 2. If the data is altered, the hash value will also change. This increases the operational efficiency and security of the transaction.

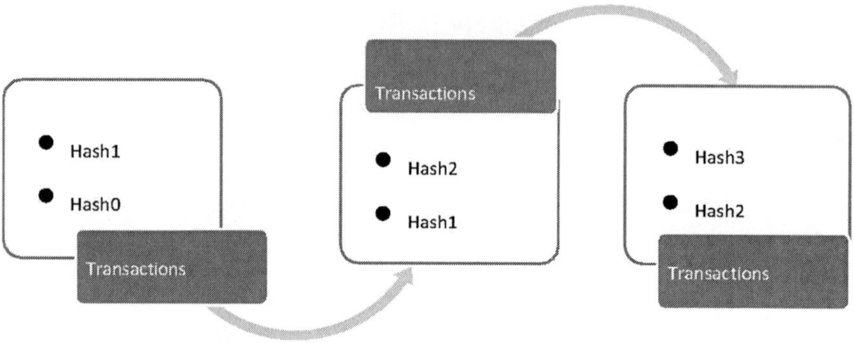

Figure 2. Hash Generation

Features of Blockchain

With the rapid development of technologies, blockchain is required to enhance security. This has a network of nodes, a distributed database system, cryptography, and a shared ledger. Blockchain could be public, private, or consortium-based, depending on the requirement. The hash function is changed, even if one piece of data is tampered with. The blockchain is immutable. Privacy and security are maintained in blockchain. User details cannot be retrieved.

Blockchain is preferred because it has a public distributed ledger that maintains all the transactions in block format. Blockchain involves end-to-end transactions. Hence, a third party is not involved in the transaction. The blockchain shares information across the nodes. This prevents fraud intervention, reduces the cost of managing the data, and reduces the time involved in transactions with augmented security. The entire blockchain could be traced back for verification and fraud detection. Blockchain records are secured by cryptography. Network participants involved in the banking transaction have a private key that acts as a digital signature. Personal information is not stored in the block. Banking transactions are verified using their digital signature. The network could be public or private. A public blockchain connects computers connected over the internet, bundles them into blocks, and adds them to the ledger. Private blockchains verify the identities and privileges of the connected users. The transactions of the users are verified by a common agreement. Blockchain validates data using mining. Mining generates a hash for each block transaction, which is tough to forge.

Blockchain-based Banking Applications

Blockchain is used in distinct arenas due to its prominent features such as distribution, decentralisation (Aarti& Vinod, 2020, 65-73), and immutability. Blockchain data is distributed among nodes. If the data is forged, it could be identified from the hash. Using decentralisation, data is protected from third-party intruders.

Apart from default banking transactions, blockchain is also used to perform other significant functionalities connected to the bank. Blockchain is a successful technology to perform financial transactions without the need for a centralised server (Saravanan, 2020, pp. 476–476). Some of them are listed below as use cases:

- Blockchain is used in **internal banking transactions, intrabank transactions** occurring across countries, confidential information exchanges between peers inside the bank, and also in ecosystems where access control is Data is shared and monetized using the blockchain.
- A bank could collect **KYC** (know your customer) using blockchain and share it across the bank so that customers do not need to provide KYC for each and every service they obtain from the bank. This could be shared with other banks through a private key.
- Blockchain has remarkable potential in **trade finance transactions,** which involve managing document transmission, importer, importer bank, exporter, exporter bank, customs, shipping, and so on. A smart contract manages document transmission, document procedure handling, verification, and validation. Transparency saves time and builds trust.
- Banks provide loans based on a lot of documents. These documents are verified by several departments. Blockchain could be used to share documents between concerned parties.

Proposed semi-centralized architecture using blockchain

In banking, if a central server fails, all the transactions fail. A hacker could hack the centralised network and tamper with the data. A semi-centralised blockchain-based bank web application was implemented using blockchain at the back-end, as shown in Figure 3. This website assists banks in providing approval for transactions without the need for third-party services. This web application eases financial transactions. Instantaneous recording and validation of user transactions is achieved using blockchain at the back end. The anonymity of the user is maintained by using a digital signature, which is unique for each transaction. On the Ethereum blockchain, transactions are performed only when certain conditions are met. These conditions are referred to as 'Smart Contracts' (Srdan & Dejan 2022, 30–42). Ether is the digital currency of Ethereum. While executing smart contracts, an initial fee needs to be paid. The user can save money by not paying extra charges to the bank for third-party services; instead, he or she works on the Ethereum network, which is very low-cost.

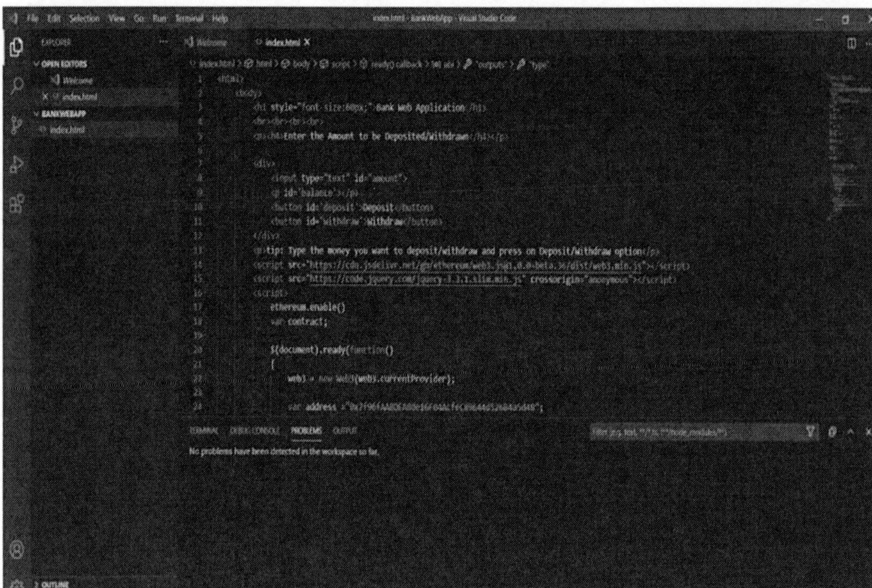

Figure 3. Website developing and interfacing with the backend

For software implementation, the following tools were used: The Visual Studio Code Remix Online IDE was used to develop, deploy, and administer smart contracts for Ethereum, like local blockchains. MetaMask was used as a crypto wallet for Ethereum, as shown in Figure 4. A Twitter account was used to get a few test emails. Ganache is an

Ethereum developer tool that simulates a local blockchain environment and tests smart contracts without connecting to any network. Node-Red is a flow-based visual programming tool used to monitor the real-time value of cryptocurrency.

Apart from the above-mentioned tools, Solidity was used to write smart contracts in the backend; Hypertext Markup Language (HTML) was used to structure the frontend; JavaScript was used to make the front-end more responsive and interactive; JSON was used to store and obtain data from the backend; Web3 describes the future internet built on blockchain; the Ethereum JavaScript API Web3 allows clients to interact with the Ethereum blockchain.

Website Design

Firstly, the Visual Studio Code IDE was installed for visualising the changes on the live website. A folder named 'BankWebApp" was created. An HTTP server was started by typing http-server in the terminal. In the URL of the localhost, an empty website was visualised. Here, a customised website was created for the banking application. A new.html file was created in Visual Studio Code inside the BankWebApp. Using HTML, the initial few lines were coded to create the input text field; buttons were created to deposit, withdraw, and show the balance.A static website was created for the banking transaction, entitled 'Bank Web Application'. Buttons used 'jQuery' and other operations to connect to the Ethereum network and perform the required transactions. This was done using the Web3 library, which is an API between the Ethereum network and the web application. This is added using a script tag. When the user loads the page, by enabling 'ethereum.enable()', he can see the balance in his account.

The Web3 library was used to connect to the front-end website and the Ethereum blockchain in the backend. The required functions need to be specified to obtain one's balance and to deposit or withdraw a given amount from the account. The front end is ready for display. This is also interfaced with the Ethereum blockchain to visualise the workings of the bank web application.

Banking Transaction

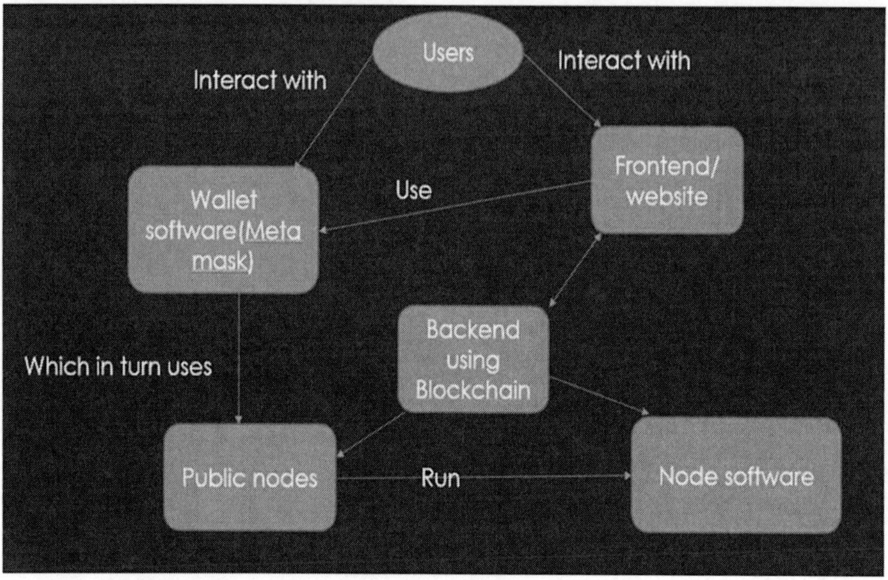

Figure 4. Process flow of the banking transaction

Figure 4 indicates the banking transaction made by the user using Ethereum. To open a new account, he needs to pay a minimum of $1 to satisfy the minimum balance requirement criteria. The bank (backend) deploys the smart contract, and the user gets the new account number, which he has to remember to do further transactions. Once the bank deploys the smart contract, an account is created. But for the user to perform any further transactions, he has to connect to the Ethereum network (something like a centralised internet), on which all the transactions are performed. Transactions are performed with the help of cryptocurrency. Ethereum is preferred in this case because of its feasibility (the current market price of various available cryptocurrencies was obtained from the node-red dashboard using the Binance node). The user could use the dashboard to monitor the price of the selected cryptocurrency. The use of cryptocurrency as a transaction fee ensures security using a private asymmetric key, as provided by the wallet software. MetaMask wallet software is chosen because it is used as the gateway to connect to the Ethereum network.

The bank provides the website to the user, which runs on the local server, and hence, no password is needed to ensure security (security is already ensured through wallet software). The user can deposit or withdraw money by simply typing the amount he wants to deposit or withdraw in the text box and pressing the deposit or withdraw button. The user will get the transaction details along with his or her block hash. Each and every transaction is stored safely in the block. The user has a record of his transactions performed in JVM. In this way, no third-party validation of the transaction is required. The transaction is done almost instantaneously, whereas in the case of normal bank transactions, an OTP is generated. This is time-consuming and susceptible to threats.

Results and Conclusion

In the front-end, initially, after the new account is opened, the balance becomes 1, as shown in Figure 5.

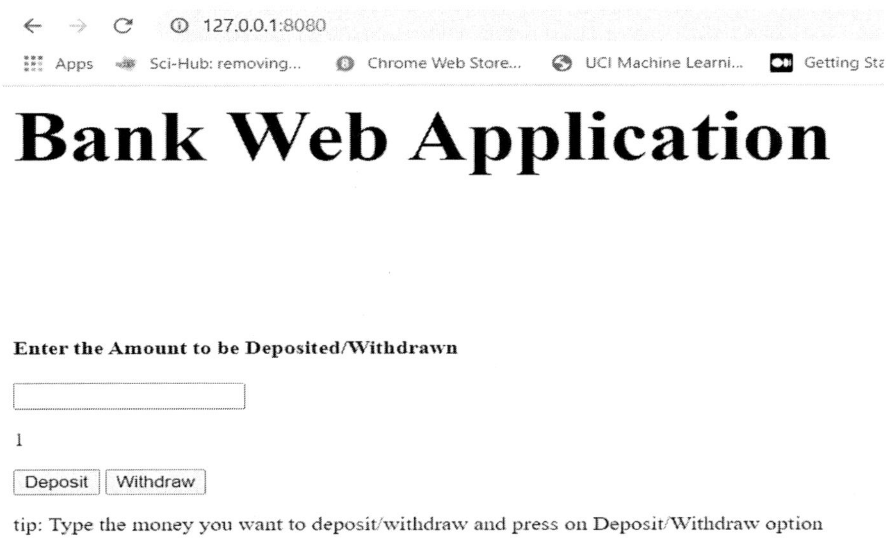

Figure 5. Front-end of the Bank Application

If the user wants to deposit 10,000 dollars or rupees, then it is updated as shown in Figure 6.

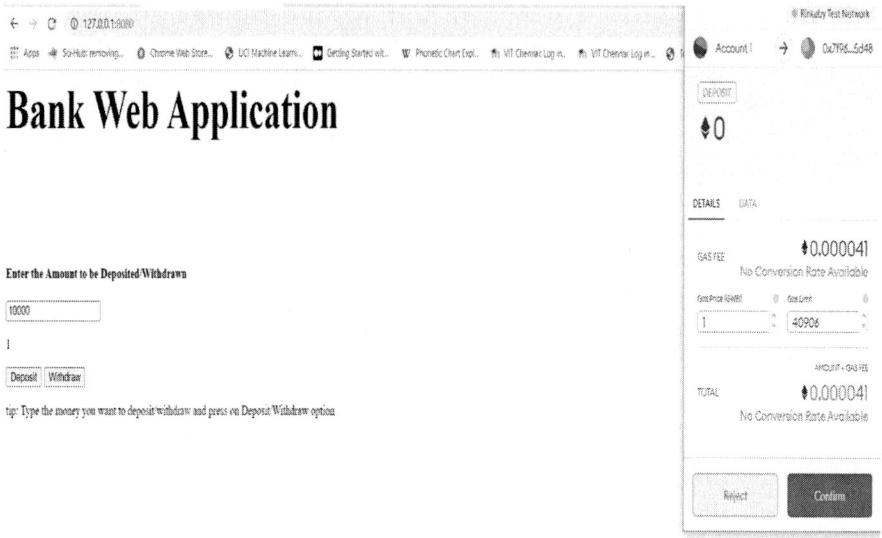

Figure 6. Banking transaction showing 'deposit' status

Once it is deposited, the wallet software opens and asks for confirmation. On the test network, only test ethernets were used. Transaction details were displayed with the hash function, as shown in Figure 7.


```
                                                              (index):96
{blockHash: "0x96b1eac461c8a97ef7ab437c34c12c72b3f5b9d438acffcd7de12fbd4660
0cab", blockNumber: 7502550, contractAddress: null, cumulativeGasUsed: 1943
106, from: "0xb5b24622345abb4a94084c51b283ccae6dd4d837", …}
```


Figure 7. Hash function in transaction details

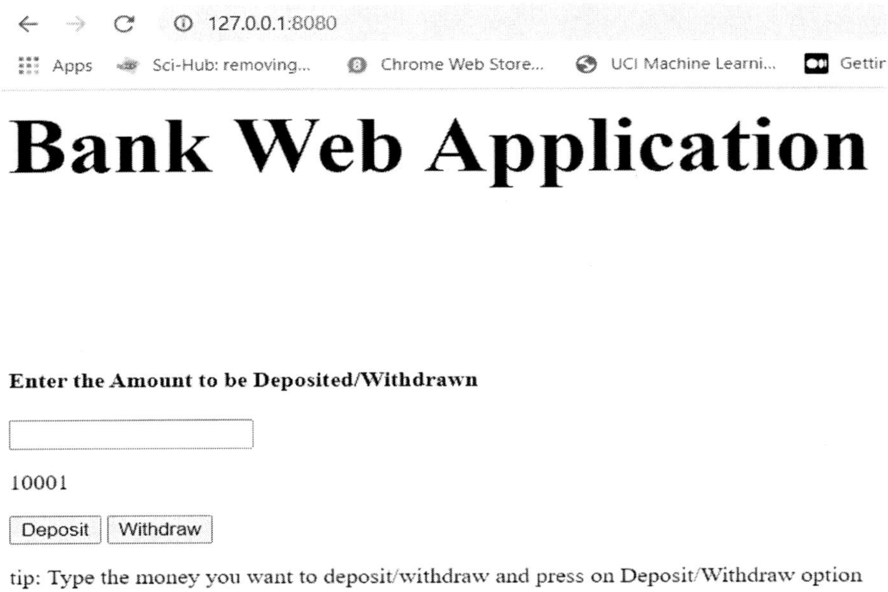

Figure 8. Updated transaction details in webpage

The web page, as shown in Figure 8, displays the updated current balance of 10001. On the back end, the bank notes the transaction results. Each activity done on the front end is reflected in the bank record, as shown in Figure 9.

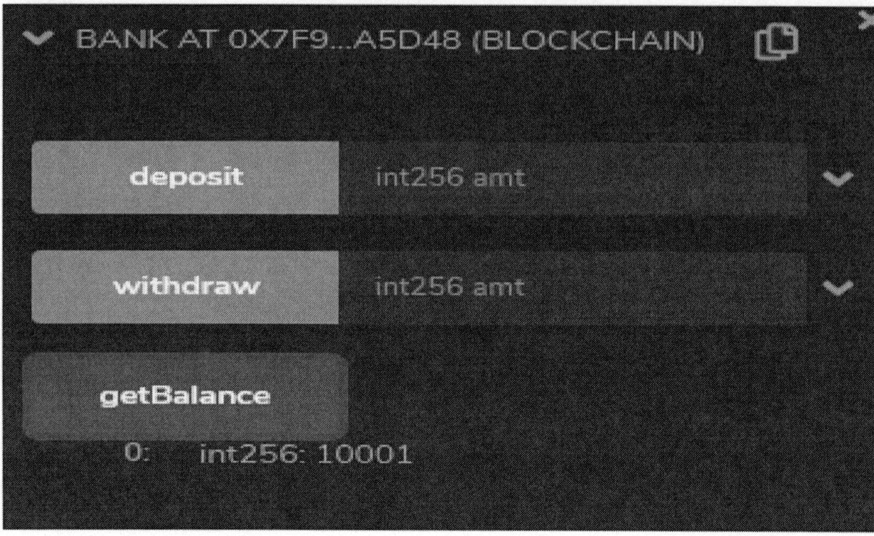

Figure 9. Bank record

Now if the user wants to withdraw $5,000 from his account, a different hash is generated, allocated, and stored along with the new block, as shown in Figure 10.

(index):96
{blockHash: "0x62c9126ccb9cfad8d2c82b970ad7b24f97db7d6797b332ff4e5b84cd6fec
▶ 505b", blockNumber: 7502573, contractAddress: null, cumulativeGasUsed: 8415
86, from: "0xb5b24622345abb4a94084c51b283ccae6dd4d837", …}

Figure 10. New Hash for the next transaction

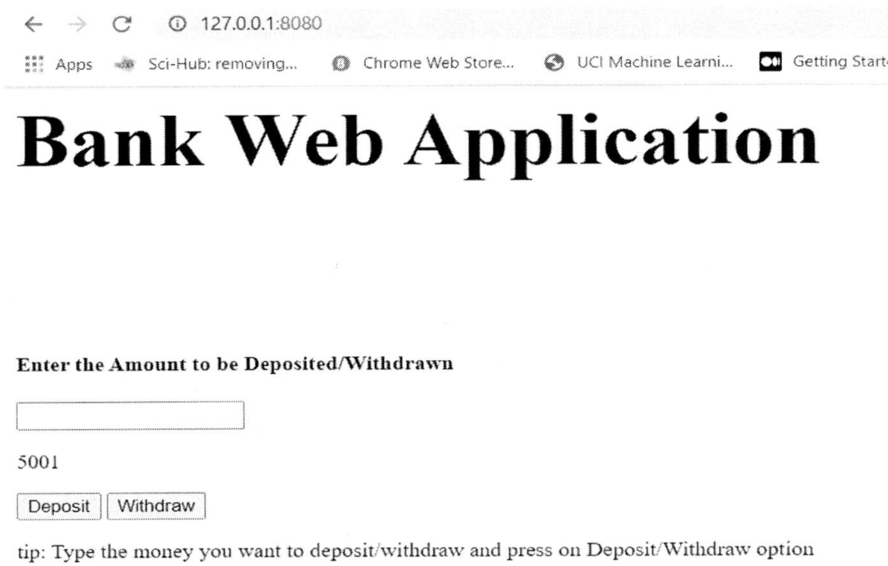

Figure 11. Webpage displaying 'Withdrawal'

The webpage, as shown in Figure 11, displays the current value of the amount. Now the current balance is 10001-5000 = 5001. From the bank side, it is verified and updated as shown in Figure 12.

Figure 12. Bank Record

Each transaction, irrespective of the bank, is performed on the Ethereum network. The back-end was visualised to check the functionality of the processes involved in the web banking application transaction.

If this application is deployed in reality, wherein each bank has its own cryptocurrency (which in turn has its own market value), and if the user is allowed to perform his own transactions, then the user has every right to know the current market value of the cryptocurrency he is paying as a transaction fee to perform a transaction. This is displayed using the Node-Red dashboard by using the Ethereum node, as shown in Figure 13. Figure 14 displays the values of the existing cryptocurrencies.

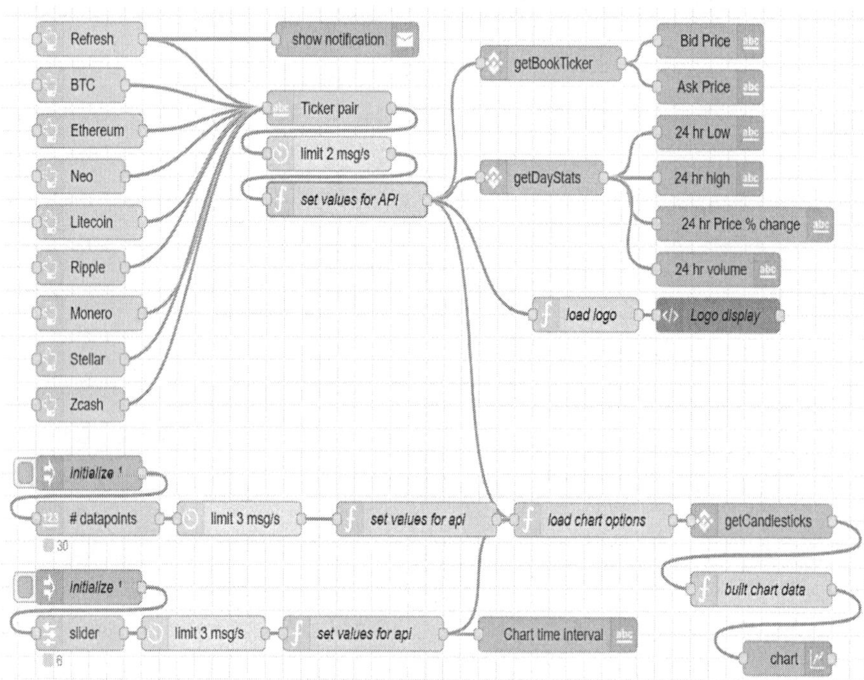

Figure 13. Node-Red with Ethereum node

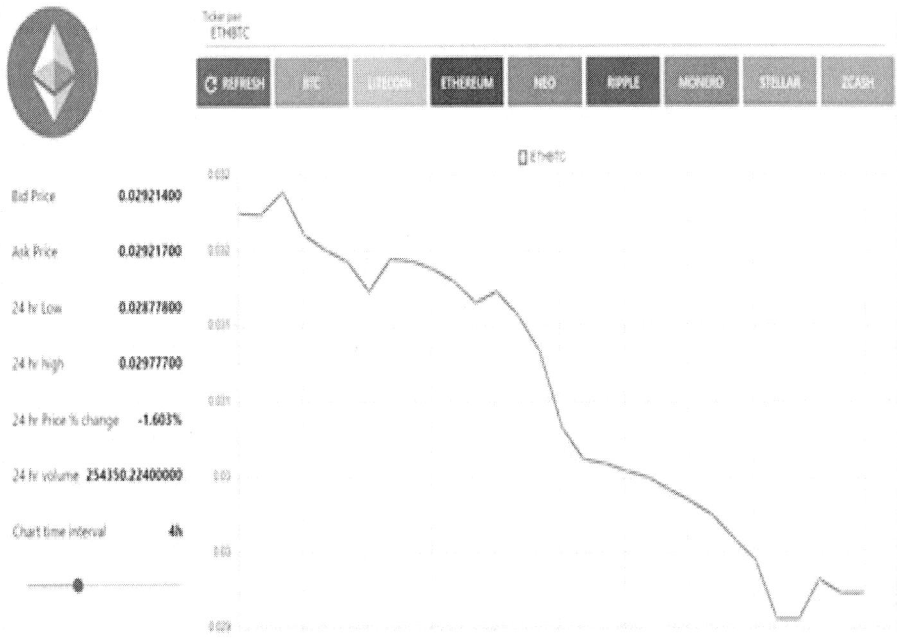

Figure 14. Existing Cryptocurrencies

More trending use cases relating to blockchain

- **Supply chain management** is transparent and scalable, with end-to-end visibility. Blockchain with IoT (Taylor et al., 2019, 4) improves the reliability and traceability of the data and the network. This deploys temperature sensors, motion sensors, vehicle information, and GPS to track shipments. In this case, sensor information is stored so stakeholders can get real-time access to the data.
- In the case of **smart cities**, sensitive information could be obtained from voice recognition, facial recognition, and so on. The data is on the blockchain. Data access is provided with immutability to the stakeholders. Identities are shared using blockchain (Christian et al., 2021).
- Digitization has improved **the automotive industry.** This is implemented using blockchain and industrial IoT to improve automation, such as in autonomous cars, automated traffic control,

and so on. With the surge in IoT, tamper-proof databases are needed (Singh & Singh, 2016).

- A patent registration and trading system was implemented using Jian et al. (2022).

Integration of AI and ML for Banking

AI, ML, and deep learning could be integrated with blockchain in the banking sector. Data analytics plays a major role in the banking sector. AI manages data access. Banking firms combine AI and ML for risk management and fraud detection. AI and ML secure data collection and processing; analyse behaviour patterns and make predictions; automate data auditing; provide personalised customer satisfaction; ensure return on investment; and much more. AI-driven apps interpret voice and facial recognition. AI-driven chatbots respond to customer queries. AI mimics humans by using its algorithms in problem solving and decision-making. AI integrated with blockchain improves trust and data integrity. In healthcare, blockchain secures data, and AI obtains insights into patient data. In the pharmaceutical industry, blockchain and AI add visibility, data integrity, and traceability. In financial services, loan processes and transaction speeds are accelerated.

Conclusion and Future Work

Blockchain applications and features are on the rise due to increased research and development in this arena. Blockchain encourages a sharing economy, robustness, immutability, flexibility, security, privacy, and so on. Banks build trust and transparency by using blockchain. However, blockchain is not functional with central bank digital currency (Natalia et al., 2020, 10; Tao & Zhigang, 2022, 265). This could be improved in the future. In the proposed research, Ethereum is preferred over other cryptocurrencies like Siacoin, Bitcoin, etc. due to its smart contract features. AI and blockchain shield applications against cyber attacks.AI identifies patterns in data behaviour, and blockchain removes fraudulent data sets. This combination extends to health record sharing, smart contracts, media sharing, and many more, apart from financial security. Blockchain is expensive. Tools could be implemented to reduce the complexity in the future. In the future, AI could be used with blockchain to improve security, privacy, risk management, and make insightful decisions (Sathish et al., 2022).

References

1. Aarti Patki & Vinod Sople, [2021]," Indian banking sector: blockchain implementation, challenges, and way forward", Journal of Banking and Financial Technology, Springer, Vol. 4, 65–73.
2. Christian Espositoa, Massimo Ficco, and Brij Bhooshan Gupta [2021], "Blockchain-based authentication and authorization for smart city applications", Information Processing and Management, 58, Issue. 2, ScienceDirect.
3. Chuan Chen, Lei Zhang Yihao Lei, Tianchi Liao, Siran Zhao, Zibin Zheng, Huawei Huang, and Jiajing Wu [2022], "When Digital Economy Meets 0: Applications and Challenges", in *IEEE Open Journal of the Computer Society*, Vol. 3, pp. 233-245, 2022, doi: 10.1109/OJCS.2022.3217565.
4. Jayasudha Subburaj, Ragavi, P. Keerthana, & C. Soundarya Veni [2020]," Block Chain Technology: An Outline", Lecture Notes in Electrical Engineering book series, LNEE Vol. 601, ICDSMLA, pp. 471-481.
5. Jian Hu, Peng Zhua, Yong Qi, QinGyun Zhu, and Xiaotong Li [2022]," A patent registration and trading system based on blockchain", Expert Systems with Applications, ScienceDirect, 201.
6. Natalia Dashkevich, Steve Counsell, and Giuseppe Destefani [2020]," Blockchain Application for Central Banks: A Systematic Mapping Study", IEEE Access, pp. 2–35.
7. J. Taylor, Tooska Dargahi, Ali Dehghantanha, Reza M. Parizic, Kim-Kwang Raymond Chood [2020]: "A systematic literature review of blockchain cyber security", Digital Communications and Networks, Vol. 6, Vol,6, No.2, pp.147- No. 2, pp. 147–156.
8. Saravanan Krishnan, Valentina Balas, Julie Golden, Y. Robinson, S. Balaji, and Raghvendra Kumar [2020]," Handbook of Research on Blockchain Technology", Academic Press,
9. Sathish Kumar, Lim, W.M., Sivarajah, U., and Jaspreet Kaur [2022], "Artificial Intelligence and Blockchain Integration in Business: Trends from a Bibliometric-Content Analysis", Information Systems Frontiers, Springer
10. Singh and N. Singh, [2016], "Blockchain: Future of Financial and Cyber Security,", *2nd International Conference on Contemporary Computing and Informatics (IC3I)*, Greater Noida, India, pp. 463–466, doi: 10.1109/IC3I.2016.7918009.

11. Srdan Tripkovic and Dejan Simic, [2022]," Using Ethereum Smart Contracts for Payment Transactions,", Lecture Notes in Networks and Systems Book Series, Springer, 562, pp. 30–42.
12. Tao Zhang and Zhigang Huang [2022]," Banking and Central Bank Digital Currency,", Korean Institute of Communications and Information Sciences, ScienceDirect, pp. 264–270.
13. Thulya Palihapitiya, [2020] " Blockchain in the Banking Industry", in Conference on Distributed Computing, Sri Lanka.

CHAPTER 7

SMART AND SECURE HEATH CARE MONITORING SYSTEM USING IOT AND BLOCKCHAIN TECHNOLOGY

PADMAVATHY T V[1], NARMADHA T V[2]

[1]SCHOOL OF COMPUTER SCIENCE AND ENGINEERING, VELLORE INSTITUTE OF TECHNOLOGY CHENNAI

[2] DEPARTMENT OF ELECTRICAL AND ELECTRONICS ENGINEERING, ST. JOSEPH'S COLLEGE OF ENGINEERING CHENNAI

Abstract

Wireless networks of the fifth generation will currently dominate transmission. Technological developments offer improved assistance through the network interface between the devices and the network. Additionally, 5G offers more advantages in terms of data security, decentralisation, and information sharing between various network providers. In 5G networks, a huge number of Internet of Things (IoT) devices are connected. These devices are strengthened by higher rates, making quick processing possible. A notable breakthrough in healthcare is blockchain, which is used for disease prediction and intelligent treatment. There may be further advantages if this blockchain technology is extended to the IoT. Because there are many connected IoT devices and more sensitive data is being shared, security is a major problem with 5G innovation.

Keywords: Blockchain; Internet of Things (IoT); 5G Networks; Digital Health Records; Decentralization

1. Introduction

As blockchain technology gained popularity after the Bitcoin whitepaper was published in October 2008, As Bitcoin's underlying technology, blockchain's primary benefit is that it enables the sharing of electronic coins across a decentralized network without the requirement for a centralized, reliable third party. Traditionally, agreements involving the exchange of virtual currency between individuals or corporations have been handled by a trustworthy third party (TTP), such as a bank. It is not advisable to rely on a TTP for a number of reasons. A trustworthy third party may malfunction, fail, or be maliciously compromised in order to make the financial system inaccessible or unsure; thus, a TTP is potentially a single point of failure that undermines a system. A TTP also imposes transaction fees and delays on transactions. The intention behind Bitcoins is therefore to overcome these restrictions associated with electronic transactions that rely on TTP. Following the release of the Bitcoin white paper in October 2008, blockchain just became a distributed ledger technique. As Bitcoin's underlying technology, blockchain's primary benefit is that it enables the transfer of electronic coins across a decentralized system without the necessity of a centralized, trusted third party. Like a bank, a trusted third party (TTP) has traditionally mediated exchanges involving digital currency transfers between individuals or businesses. Depending on a TTP is undesirable for a variety of reasons. A trustworthy third party may malfunction, fail, or be maliciously compromised in order to make the financial system inaccessible or unsure; thus, a TTP is potentially a single point of failure that undermines a system. A TTP also imposes transaction fees and delays on transactions. The intention behind Bitcoins is therefore to overcome these restrictions associated with electronic transactions that rely on TTP.

In recent decades, the advancement of integrated systems has followed an ascending curve. In fact, embedded systems have penetrated such domains as Industry 4.0, agricultural production, data management, commerce, education, and health. Due to their importance to human health and well-being, the use of technological advances in these fields has increased recently. Using algorithms for artificial intelligence to process data from various sensors is a highly effective new method for detecting anomalies, diseases, problems, and alerts. Indeed, it enables an effective medical diagnostic assistance solution based on embedded systems.

It is also important to note that the purpose of this chapter is not only to identify the use cases or examples of blockchain-based technology in healthcare but also to understand the limitations and challenges for blockchain-based smart healthcare as well as the recent trends in terms of the technical approaches, methodologies, and concepts used in establishing these applications (and in overcoming the limitations) in order to identify the potential areas for future research. In addition, this chapter reviews many materials that were not released at the time of the previous reports. As previously mentioned, the implementation of blockchain technology in healthcare is a comparatively recent framework that is expanding rapidly; consequently, there are numerous new articles on the subject.

2. Background

The most significant benefit of smart and secure healthcare using advanced technology like IoT and blockchain-based systems is that it is extremely convenient. A more specific application of smart and secure healthcare is the use of smartphones to notify users about behavioral healthcare services [Olla 2015, 299–308]. In addition, it enables patients and healthcare professionals to communicate in real time without physically meeting. The blockchain provides patients and physicians with secure data communication [Patel2018, 324–336]. IoT is recognized as a transformative digital medium that played a significant role in the creation of next-generation technology initiatives such as smart cities, building automation, and the smart medical system [Alam 2020, 17–25]. Similar to an essential component of remote health monitoring, an interrelated medical system consists of a variety of embedded or transported biosensors on the patient. This is used to collect information about a patient's health and transmit it to medical applications utilizing wireless networking.

Increasing numbers of IoT devices are now connected and automatically transmit data through the communication channel. This results in insecure data sharing, data loss, connectivity issues, etc. 5G technology fixes connectivity problems. Adaptability, reliability, bandwidth allocation, identity verification, identification, lack of standards, and interconnectivity are a few 5G IoT architecture issues. Blockchain technology addresses a few security issues. In 5G networks, the blockchain has indeed been integrated with more IoT networks in order to offer more new opportunities for assistance and applications [Alam 2020, 46–58]. This section discussed blockchain technology for 5G IoT systems.

In recent years, blockchains have emerged as an innovation that may have taken over the globe [Tanweer Alam 2020, 108–115; Alal 2020, 28–41].

Blockchains are nothing more than a common ledger that keeps track of system-wide transactions and events [Tanweer Alam 2019]. The most crucial aspect of a blockchain is that when a piece of data is added to the shared database, it cannot be altered [Tanweer Alam 2020, 543-558]. Any storage of data on the ledger seems entirely secure. In exchange for a person making a change to a block, he or she is required to make alterations to all subsequent blocks [Mohamed Benaida 2020, 2164–2174; Tanweer Alam 2018, 3378–3387]. The Internet of Things (IoT) is a contemporary and evolving idea [Alam 2018, 450–456; Mohammed Aljohani 2015, 1322–1327; Tanweer Alam2015, 371–375] that provides online communication using sensing devices for intelligent proof of identity and management in such a different communications environment. This emerging aspect, from the perspective of independent living, would enable new interaction routes between disabled patients and care facilities via innovative network security interfaces. Telehealth has been defined as "innovations in wireless technology, health instruments, and telecommunications for healthcare." The above innovation provides mobility and an always-connected user experience for public healthcare applications [Mohammed Aljohani 2016, 1-6; Tanweer Alam2017, 86–94; Alam 2017, 4526–4538]. Metadata has become an indispensable component of health care. In terms of information transmission, patient management, and analysis, the digitalization of health data has also produced numerous benefits. The quantity of platforms producing health information has increased exponentially, including within health organizations.

Blockchain technology could have a substantial positive impact on medical services [Tanweer Alam 2019, 342–359; Peng Zhang 2018, 445–456]. Using the distributed ledger to protect metadata from anomaly-based tampering attacks [Andreas Hoglund 2017, 16–22; Rababah 2019, 1–10], the blockchain-based method was devised by Zhang [2019, 124–138] to optimize physiological information security while also improving transmitting data performance.

Information security is an increasing difficulty. The role of blockchain in this study is to manage health information interconnectivity and security issues, such as enabling effective, approved collaboration between patients and social insurance providers (specialists, insurance agencies) and transmitting persistent data securely to a variety of organizations and devices. Healthcare information ought to be more private and confidential. This information should be protected from alterations and deletions of stored data. If an unauthorized person gains access to the data, it will either be sold or misused. In 5G networks, the size of healthcare data grows, necessitating the implementation of additional security measures. Using blockchain techniques, additional algorithms have been proposed for securing health data. A key management algorithm that uses a novel approach to secure healthcare data is presented in this chapter.

3. Blockchain Technology in the Medical and Health Sciences

This study aims to extract and analyse the blockchain in order to improve processes and services in health coverage, medical sciences, and biomedical research. Digital medical files and individual medical records are the most endorsed areas for blockchain, according to the results. Permissions, integration, provenance, and integrity of the information are the challenges that blockchain technology is intended to address in this field. The research indicates that efforts are underway to implement blockchain technology in the healthcare industry.

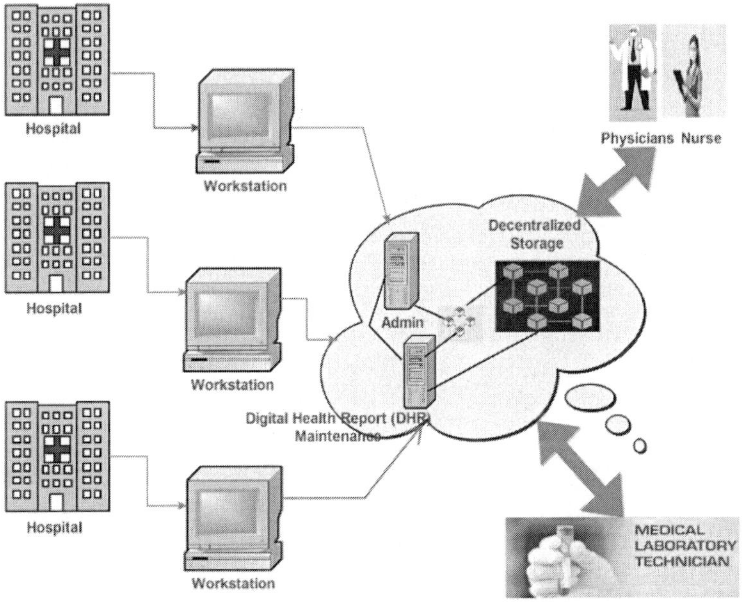

Figure 1: Application of Blockchain in Health Industry

3.1. Use Cases of Blockchain Technology in the Healthcare Industry

In this chapter, just a few application cases for blockchain in healthcare are identified. They are

- Digital Health Records (DHR)
- Clinical trials or studies
- Supply chain for pharmaceuticals
- Health Insurance
- Remote observation using IoT gadgets

3.1.1. Digital Medical Records (DMRs)

The DMR is a computerized recording of a patient's medical history. It is a file with essential information such as a diagnosis, prognosis, and lab investigation findings. Updating medical records is a vital aspect of enhancing the intelligence of the healthcare industry as a whole [Zehui Xiong 2018, 33–39]. Global healthcare systems are battling with the issue of data silos, which affect both patients and providers. Their medical history is incomplete, and their healthcare practitioners are unaware of this. Integrating blockchain technology with DMR, IoT, and 5G technology can solve this problem and facilitate healthcare data interoperability [Peng Zhang 2018, 445–456; Zainab Alhadhrami 2017, 1-4]. Permissionless or public blockchains, on the other hand, are unsuitable for such applications since any anonymous network participant can access the data. In many nations, patient information is often surrounded by different privacy rules. Everyone on the blockchain network can see the data. This can be avoided by adopting a privatized blockchain system instead of a public blockchain. In a personal blockchain, a private blockchain network requires permission to join and view the data, while a public blockchain network is open to anyone with internet access. This sort of network can pose a number of security risks involving privacy. This could potentially transform the healthcare business, which is now dependent on psychiatric hospitals. This allows patients to make more informed decisions concerning their healthcare treatment.

With the use of a program called Intelligent contracts, which are smartly written lines of code, blockchain permits the patient to have authority over the access protocols and determine who to grant access to. When a predetermined event occurs, these codes are run. A series of conditions have been met. These are lines of code that run on their own. Because the data in these DMRs is encrypted with cryptographic hash algorithms. It is economically feasible to validate the transaction's integrity on the blockchain. There is only one correct interpretation of the records with the patient and the medical professionals, which is a benefit of combining blockchain with DMRs.

3.1.2. Clinical Trials or Study

Clinical trials or research would be another application for blockchain. A difficult feature of any medical trial is finding patients to participate in clinical trials [Tanweer Alam 2018, 97–111; Yan Zhuang 2019, 1276–1285]. The sponsor, the patient, and the main investigator all have concerns that need to be addressed, which is why the majority of clinical studies fail to fulfill their recruitment targets on time. Conducting clinical trials with insufficient enrollment reduces the strength of definitive findings or results in an early trial termination. Future clinical studies can greatly benefit from blockchain technology because it has a complete record of each participant's medical history. For instance, if a volunteer moves on with the trial while having an allergy to a certain kind of medical ingredient that is being utilized in the study, this could have serious consequences for the patient.

Blockchain can offer a patient's whole medical history, including any prior adverse reactions to specific pharmaceutical compounds, enabling clinical researchers to make well-informed trial decisions. Clinical trial recruitment benefits from blockchain technology's peer-to-peer decentralized system with globally monitored activities, data protection, and patient privacy. Patient enrollment, record keeping, data exchange, and data privacy are all highly delicate aspects of clinical trials, as are numerous privacy and security issues that may occur. Blockchain technology can provide protection for the intricate and extensive clinical studies that require a variety of records [Mehdi Benchoufi 2017, 1–55].

3.1.3. Supply Chain for Pharmaceuticals

Every industry faces formidable challenges, with supply chain management being particularly difficult in the healthcare sector. A broken supply chain in the healthcare industry may have an impact on patient safety. Blockchain facilitates the first step of drug manufacturing tracking. The precise ingredients can be tracked as the pharmaceuticals are manufactured thanks to blockchain technology. It offers full medicine traceability from the manufacturer to the user or patient, who is the final user. This guarantees the medications' authenticity to the appropriate parties. To address this problem, a conceptual framework [Monalisa Sahoo 2020, 213-222; Faisal Jamil 2019, 505] was put forth. The manufacturer, distributor, transporter, pharmacy, and hospital are the five departments involved. It is possible to track and keep tabs on the movement of the pharmaceuticals as they move through each department.

3.1.4. Health Insurance

Medical insurance protects you against any crippling expenses brought on by a terrible accident or sudden medical issues. Three organizations—healthcare providers, insurance companies, and clearinghouses—handle the medical claims. The dispatcher handles the dispute between the insurance companies and the healthcare providers. The clearinghouse will have power over the situation and could behave nefariously on its own. With the aid of blockchain technology, this can be prevented. Blockchain offers a systematic method where no one person has complete control, eliminating the need for clearinghouses to mediate. Insurance companies can use smart contracts.

As soon as a code is recorded on the chain, it cannot be changed, making claims difficult. This can assist patients in receiving their claims quickly and truthfully. One of the most important topics that is now being explored is IoT privacy and security. Traditional security solutions are inadequate because of the IoT's resource limitations. Wireless body area networks, or WBANs, were suggested as a blockchain solution for tracking the patient's vital signs. These essential pieces of information are obtained through gadgets like smart wearables. These wearable devices collect raw data, which is transmitted to a master device, where it is analyzed and provided as a notification to the accountable parties. Sensible contracts, when a specific set of conditions are met, are utilized to carry out specific actions. Based on the raw data collected from the WBANs, these agreements are used to carry out specific tasks, like alerting caregivers in case of an emergency. In certain situations, when the wearable gadget detects that the patient's insulin levels have fallen below the critical limit specified in the permission blockchain, it sends an alert to the centre of care so they may take the appropriate action without delay to guarantee the patients' protection, and this is maintained in the database in case the occurrence is mentioned again in the future.

3.1.5. Remote observation using IoT gadgets

The remote healthcare server allows the patient, nurse, or doctor to submit a transaction to either create a new task or receive a response from a job that has already been generated. The remote healthcare server then requests the blockchain network to carry out a task in accordance with the request. The server also transmits tasks created by the client to retrieve current vital-sign data, device data, and device status. Along with the particular patient data specified in accordance with the smart contract, the collected data, which comprises vital information and device information such as device ID, device type, and device status, is recorded in the ledger. Finally, following a successful transaction, a notification is sent to the involved party.

On a blockchain, each user often needs to sign their identity before completing a transaction in order to be recognized as a legitimate user. Each participant sends the request message to the blockchain network, including the physician, patient, and caregiver. Additionally, identity and access management, which is in charge of issuing the secret key for approving a registration procedure, obtains the registration request through an application server. The public key for the answer and the enrolment certification are generated and sent to the client. At long last, transaction data is used to create the certificate, which is then used to sign the transaction.

Only authorized users have access to and can use blockchain services after successful enrolment. Users can obtain information regarding physiological indicators using patient applications. The server receives the request and the vital-sign data and, in response, activates the smart contract associated with that transaction. The consensus procedure is then carried out in the blockchain network, appending the transactional data to the blockchain and recording the vital-sign data in the state database. After the transaction has been successfully completed, the patient is notified about the ledger upgrade. Patients can also create functions to operate on medical sensors, such as reading heart rate from an ECG sensor. Additionally, users can designate particular duties in accordance with requirements. The precise and complete request is submitted to the IoT server for healthcare, which further translates it into the sensor's prescribed protocol and sends it to the designated sensor for operation. The way the target sensor acts, vital-sign data is returned in accordance with the request to the blockchain network and healthcare IoT server. The user is shown this vital-sign data in the server application in the form of results.

3.2. 5G with Block-Chain and IoT Technologies in Health Industry

Blockchain technology has enormous potential in one particular sector. The related field of delivering healthcare services remotely via telecommunications is called telemedicine. The use of blockchain technology by healthcare professionals to access a decentralized ledger to find the required information This blockchain and 5G technology integration also enables information blockages to be removed. Idealistically, this is the capability for individuals to view their health records when sharing the data with their doctors in a secure manner as necessary. Whenever it comes to security, there has been telemedicine. If the doctor-patient relationship is secure, the risk of a data breach is increased because the patient is indeed not secure [AnweerAlam 2018, 146–162].

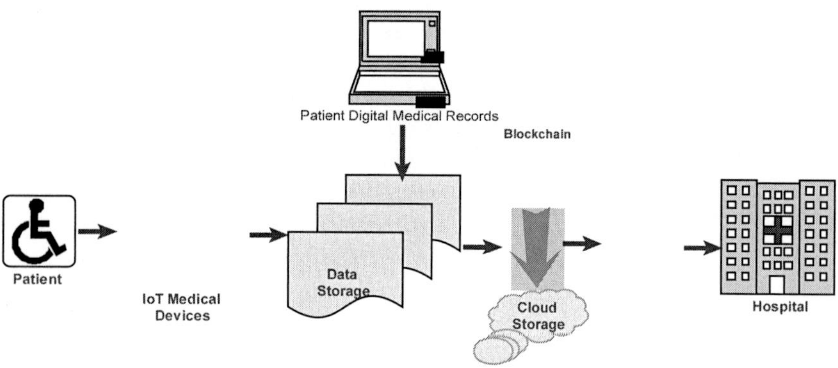

Figure 2: System Architecture

Figure 2 shows 5G, IoT, and blockchain-based healthcare systems. When combined, 5G and blockchain technology can build a safe network. Technology based on blocks also maintains the medical file in a safe, dispersed system that is data- and information-rich, allowing healthcare professionals to keep more thorough patient histories because their medical data is safely encrypted. The introduction of 5G will have many advantages, including faster data transfer, a load of files seamlessly loaded with data, real-time telemedicine visits, smart device monitoring from a distance, and the fact that 5G technology streamlines all medical procedures and eliminates the primary Internet barrier, broadband, resulting in really intelligent hospitals. Additionally, using blockchain technologies to deploy 5G will allow safe information transmission between medical professionals and patients at a quick transmission rate.

3.2.1. Processing Medical Records

Collecting data from sensors like blood pressure, body temperature, oxygen concentration, and pulse rate sensors is the initial stage in constructing this system. In order to connect, manage, and control various services in the intended system, the connection layer, application layer, and management layer are all crucial. The management layer administers cloud storage via device management, which creates channels for data exchange between hardware devices, while the connectivity layer links networking components and controls network and WiFi modules. At the top, there is an application layer that offers functions like patient monitoring, appointment scheduling, a display of body temperature, and alerts for changes in bodily parameters.

The microprocessor in IoT medical devices is connected digitally to a blood pressure sensor, a body temperature sensor, and a pulse-rate sensor. The microcontroller functions as a centralized device that receives inputs from various sensors and data that is saved in a database, implements blockchain technology to keep the data secure, and then maintains data in the cloud utilizing 5G services over the Internet.

The expert system receives the data that the server has processed after receiving it from the microcontroller. The expert system analyzes the data and then compares it to previously stored data. Depending on the specifics of the case, three possible treatment options can be proposed. An expert system will first recommend a course of treatment if it has knowledge about the ailment for which the symptoms are being experienced.

Second, if not, the information is sent to an expert, who makes a recommendation for the patient's treatment within 24 hours. When a patient has an illness for which neither the expert system nor the specialist has a record of the symptoms, the specialist consults with a group of medical professionals. After holding meetings to discuss the condition, conduct trials, and gather all pertinent data, the doctors' panel will prescribe a course of therapy to patients either through the hospital's healthcare center or through an online expert system. In order to keep up with emerging diseases and their therapies, the expert system is continually updated.

3.2.2 Cloud Storage Protection Mechanism

Three different categories of data that support decision-making were included in the suggested framework. The data from the IoT-based sensor application was included in the three types of data. These three types of data have been maintained in the database and utilized for further processing and analysis. The dataset as well as the medical records were collected from the clinical pharmacist based on the severity of the disease. The cloud platform provides ample room for storing a lot of data. The goal of the suggested security system is to protect healthcare data effectively. A crucial element of the healthcare system is the security mechanism. Blockchain technology, which may maintain an indestructible ledger to store medical files and enable patients and healthcare institutions to handle information transparently, is one such technology that is used to safeguard the data of patient medical records.

Blockchain technology is used in the suggested system to provide the clinical decision support security mechanism. Patient records saved in the database are subjected to blockchain technology, and data is divided into distinct blocks to ensure data transparency. Using 5G technology's high transmission rate, secure data is kept in online storage and sent to healthcare providers.

3.2.3. Clinical Decision Making System (CDMS)

A Clinical Decision Making System (CDMS) is a piece of software that analyzes data and aids healthcare professionals in making decisions and supporting patients' health improvements. CDMS allows for integrated processes to offer recommendations for care plans and help when administering care. The CDMS focuses on leveraging knowledge management to obtain clinical advice and is supported by a variety of patient-related data sources. Data mining can be used with a clinical support system to examine whether a patient's medical history coincides with information important to the clinical study.

When a patient's characteristics are matched to a computerized clinical knowledge base, traditional CDMS software is used as a direct aid in clinical decision-making. Patient-specific assessments or suggestions are then offered to a clinician for consideration.

The Clinical Decision Making System (CDMS) is a key component of the healthcare system in the proposed system. In the CDMS, professionals have numerous different disease remedies stored. If the patient's symptoms and the CDMS's record match, the CDMS will recommend a course of treatment for the pertinent illness. The CDMS sends the patient's data to healthcare practitioners for advice or illness treatment if the patient's disease symptoms do not match the record that is kept in the CDMS.

3.2.3. Structure of CDMS

The following elements make up the structure of a Clinical Decision Making System (CDMS):

- Knowledge-based Environment
- Digital health records for patients
- Machine Learning
- Inference Engine
- Search Engine
- User Interaction
- Knowledge Base
- User
- Information of Patients

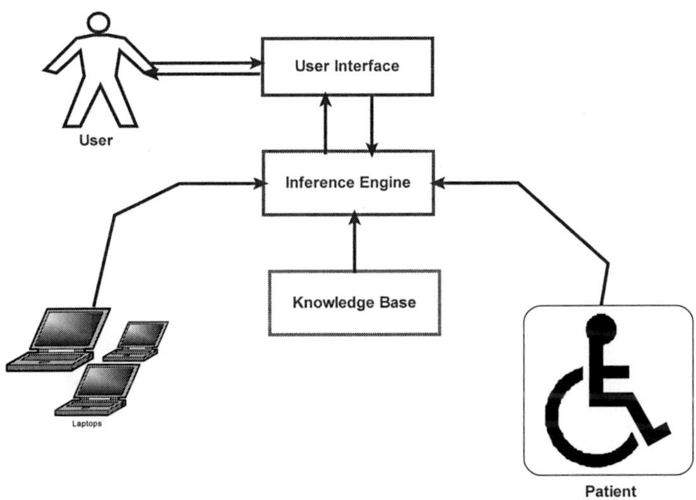

Figure 3: Structure of CDMS

Figure 3 shows the structure of the clinical decision-making system. The database contains all the data and information that the CDMS needs, and the CDMS provides users with useful information about the disease. Three types of content are possible: integrated knowledge data, organized or unstructured text data, and content that has been processed using structured data from a digital health record application.

Being a key part of the CDMS, the inference engine is useful in analyzing and interpreting knowledge. Although the inference engine has the ability to analyze patient data, patient data files are not kept there. During the process, the therapeutic inference engine activates and interprets the applicable protocol. The inference engine understands the information, such as (i) medicine dosage in connection to kidney function; (ii) electrolytes used while taking heart medication; (iii) and the option of an anticoagulant in the event of cardiac arrest.

The user interface gathers actual data, which an inference engine then processes and stores in a database before communicating to the user.

4. Conclusion

Blockchain technology, which is cutting-edge, has the potential to completely transform the healthcare industry as a whole. The characteristics of blockchain, such as decentralization, immutability, and transparency, provide an excellent foundation on which the healthcare industry may improve. Data security is provided by the fact that once registered on the blockchain, data cannot be changed or corrupted. Because blockchain is decentralized, there is no longer any need for a third party to authenticate the transactions on the chain, giving the patient full control over their own health information that was previously inaccessible due to the usc of conventional centralized databases. The optimization of electronic medical records, clinical trials, medical insurance, and remote patient monitoring via IoT devices are just a few of the many applications that this technology can be used for in the healthcare sector.

An IoT with 5G, blockchain, and a cloud-based Clinical Decision Making System (CDMS) is presented in this chapter. A framework that is employed to make predictions about the severity of the condition Medical sensors are used to collect information about patients, and earlier patient information aids in the effective care of the patient. The information obtained from various sources is kept in the database. Blockchain technology is put into practice, keeping the data safe, and then the data is kept on a cloud server. The securely stored data in the cloud using 5G storage is transmitted to healthcare providers' services. A few of the many benefits of the suggested technology in terms of healthcare services are stated below:

The proposed technology works efficiently to deliver healthcare services. This system minimizes time by overcoming the environment's limiting factors. This approach is also better suited for residents of remote areas without access to adequate medical services. The technique is also most suitable for individuals with disabilities who cannot travel vast distances. Blockchain technology has the ability to solve problems with data protection, security, sharing, and storage. Additionally, the suggested solution uses blockchain technology to guarantee the highest level of transparency in the security of medical data inside the healthcare system. Also, utilizing 5G services improves the effectiveness of healthcare services.

References

1. Alam, Anweer, Mohamed Benaida. 2018. "CICS: Cloud–Internet Communication Security Framework for the Internet of Smart Devices". International Journal of Interactive Mobile Technologies (iJIM), pp 146-162, Elsevier.
2. Alam, Tanweer, Abdulrahman A. Salem, Ahmad O. Alsharif, and Abdulaziz M. Alhejaili. 2020. "Smart Home Automation Towards the Development of Smart Cities". Journal on Computer Science and Information Technologies. No.1(1), pp. 17-25, Springer.
3. Alam, Tanweer, and Mohamed Benaida. 2020. "Blockchain, Fog and IoT Integrated Framework: Review, Architecture and Evaluation". Technology Reports of Kansai University, No.62(2), pp16-22.
4. Alam, Tanweer, and Mohammed Aljohani. 2015. "An approach to secure communication in mobile ad-hoc networks of Android devices". International Conference on Intelligent Informatics and Biomedical Sciences (ICIIBMS), IEEE Xplore. pp. 371-375.doi.org:10.1109/ICIIBMS.2015.7439466.
5. Alam, Tanweer, and Mohammed Aljohani. 2015. "Design and implementation of an Ad Hoc Network among Android smart devices". In Green Computing and Internet of Things (ICGCIoT). IEEE Xplore. pp. 1322-1327.doi.org/ 10.1109/ ICGCIoT. 2015.7380671, IEEE Xplore.
6. Alam, Tanweer, and Mohammed Aljohani.2016. "Design a new middleware for communication in ad hoc network of android smart devices". Second International Conference on Information and Communication Technology for Competitive Strategies. ACM Proceeding.pp.1-6.doi.org/10.1145/2905055.2905244.

7. Alam, Tanweer, Mohamed Benaida. 2018. "The Role of Cloud-MANET Framework in the Internet of Things (IoT)". International Journal of Online Engineering, No. 14(12), pp. 97-111, Horizon Research Publishing.
8. Alam, Tanweer, Mohamed Benaida. 2020. "Blockchain and Internet of Things in Higher Education". Universal Journal of Educational Research.pp.2164 - 2174, Horizon Research Publishing.
9. Alam, Tanweer, Rababah, B. 2019. "Convergence of MANET in Communication among Smart Devices in IoT". International Journal of Wireless and Microwave Technologies (IJWMT), No.9(2), pp.1-10, MECS Press.
10. Alam, Tanweer. 2017. "Fuzzy control based mobility framework for evaluating mobility models in MANET of smart devices". ARPN Journal of Engineering and Applied Sciences. No. 12(15), pp.4526-4538, Springer.
11. Alam, Tanweer. 2017. "Middleware Implementation in Cloud-MANET Mobility Model for Internet of Smart Devices". International Journal of Computer Science and Network Security, No.17(5), pp. 86-94, Science Direct.
12. Alam, Tanweer. 2018. "A Reliable Communication Framework and Its Use in Internet of Things (IoT)". International Journal of Scientific Research in Computer Science, Engineering and Information Technology (IJSRCSEIT).No.3(5),No.3(5),pp.450-456, Springer.
13. Alam, Tanweer. 2018. "A reliable framework for communication in internet of smart devices using IEEE 802.15.4". ARPN Journal of Engineering and Applied Sciences, No.13(10), pp.3378-3387, IEEE.
14. Alam, Tanweer. 2019. "Blockchain and its Role in the Internet of Things (IoT)". International Journal of Scientific Research in Computer Science, Engineering and Information Technology, pp 342-359, Springer.
15. Alam, Tanweer. 2019. "IoT-Fog: A Communication Framework using Blockchain in the Internet of Things". International Journal of Recent Technology and Engineering (IJRTE), No.7(6), Blue Eyes Intelligence Engineering and Sciences Publication.
16. Alam, Tanweer. 2020. "Cloud Computing and Its Role in the Information Technology". IAIC Transactions on Sustainable Digital Innovation. (ITSDI), pp. 108-115, APTIKOM.

17. Alam, Tanweer. 2020. "Design a blockchain-based middleware layer in the Internet of Things Architecture, JOIV: International Journal on Informatics Visualization". No.4(1). pp. 28-31, Springer.
18. Alam, Tanweer. 2020. "Efficient and Secure Data Transmission Approach in Cloud-MANET-IoT integrated Framework". Journal of Telecommunication, Electronic and Computer Engineering (JTEC). No. 12(1).
19. Alam, Tanweer. 2020. Internet of Things: A Secure Cloud-Based MANET Mobility Model. International Journal of Network Security.No.22(3), pp 543-558, IEEE.
20. Alhadhrami, Zainab, Salma Alghfeli, Mariam Alghfeli, Juhar Ahmed Abedlla, and Khaled Shuaib. 2017. "Introducing blockchains for healthcare" International conference on electrical and computing technologies and applications (ICECTA). IEEE Xplore. pp.1–4. doi.org:10.1109/ICECTA.2017.8252043.
21. Benchoufi, Mehdi, and Philippe Ravaud. 2017. "Blockchain technology for improving Clinical Research Quality". Benchoufi and Ravaud Trials.No.18(1), pp.1-5, National Laboratory of Medicine, Pub.Med.
22. Hoglund, Andreas, Xingqin Lin, Olof Liberg, Ali Behravan, Emre A. Yavuz. 2017. "Overview of 3GPP release 14 enhanced NB-IoT", IEEE Network. No.31(6), pp. 16-22, IEEE.
23. Jamil, Faisal, Lei Hang, Kyu Hyung Kim, and Do Hyeun Kim. 2019. "A novel medical blockchain model for drug supply chain integrity management in a smart hospital". Electronics.No.8(5). pp.505, MDPI.
24. Olla, Phillip, and Caley Shimskey. 2015. "Health taxonomy: a literature survey of mobile health applications". Health and Technology, pp.299-308, Springer.
25. Patel, Vishal. 2018. "A framework for secure and decentralized sharing of medical imaging data via blockchain consensus". Health informatics journal. pp. 324-336, Springer.
26. Sahoo, Monalisa, Sunil Samanta Singhar, and Sony Snigdha Sahoo. 2020. "A blockchain based model to eliminate drug counterfeiting". In Machine Learning and Information Processing. Springer. pp. 213–222.
27. Xiong, Zehui, Yang Zhang, Dusit Niyato, Ping Wang, and Zhu Han. 2018. "When mobile blockchain meets edge computing". IEEE Communications Magazine.No.56(8). pp.33–39.

28. Zhang, Peng, Douglas C Schmidt, Jules. 2018. "White, and Gunther Lenz. Blockchain technology use cases in healthcare". In Advances in computers. No.1(41). pp-445-456, Elsevier.
29. Zhang, Rui, Rui Xue, and Ling Liu. 2019. "Security and Privacy on Blockchain". ACM Computing Surveys (CSUR), ACM, pp 124-138.
30. Zhuang, Yan, Lincoln R Sheets, Zonyin Shae, Yin-Wu Chen, Jeffrey JP Tsai, and Chi-Ren Shyu. 2019. "Applying blockchain technology to enhance clinical trial recruitment". In AMIA Annual Symposium Proceedings. American Medical Informatics Association. pp.1276-1285, National Laboratory of Medicine, Pub.Med.

CHAPTER 8

EXPLOITING VULNERABILITIES OF LINUX OPERATING SYSTEM

SUBBULAKSHMI T, JAYASUDHA M

SCHOOL OF COMPUTER SCIENCE AND ENGINEERING, VELLORE INSTITUTE OF TECHNOLOGY, CHENNAI, INDIA

Abstract

A rootkit is a suite of one or more programs that allows a third party to hide files and activities from the administrator of a computer system. An intruder takes advantage of one or more known vulnerabilities on a particular computing platform to deliver and install the rootkit. Once the rootkit is in place, the intruder can use the infected system while remaining undetected. This chapter aims to develop a rootkit that has different functionalities, including the hiding and unhiding of TCP and UDP ports and process hiding and unhiding. It hides the processes from the system, hence thrashing the operating system. Upon insertion of the key sniffer module into the kernel, a copy of each and every key that is entered into the terminal is created. Prevention mechanisms like detection and prevention for the exploitations caused by the rootkit are created. The goal of this chapter is not just the detection of kernel rootkits but also to exploit the system and gather as much data from it as possible.

Keywords: port hiding and unhiding, thrashing, key-sniffer, process hiding and unhiding, root privileges.

Introduction

Rootkit is likely one of the most confusing terms used in today's security arena. Although the name rootkit suggests a component that allows obtaining root access to a computer system, its only purpose is to help an attacker preserve previously obtained root access. A rootkit is a collection of programmes that a hacker uses to mask intrusions and obtain administrator-level access to a computer or computer network. The goal of the rootkit is to make the activities of the hacker as invisible as possible and avoid detection by the system administrator. This way, the attacker can remain hidden on the system for long periods of time.

To accomplish this goal, the attacker must have the highest levels of privilege in the system, so the user has to have previously obtained root access to the system by escalating privileges or accessing it remotely through a root exploit. A well-designed rootkit can hide files, data, processes, and network ports and typically survive a system restart. This stealthy design allows the rootkit to perform malicious activities such as keystroke logging or give a remote attacker control of the infected system. In spite of current rootkits being extremely stealthy, there are still a number of techniques that have been developed to detect their presence. A detection programme is a shell script that checks system binaries for rootkit modifications. It can also detect Loadable Kernel Modules (LKM) rootkit-configured files. It also checks for hidden processes by checking the output of the 'ps' command in the 'proc' directory. Many rootkits run as hidden processes, so the detection programme detects them and also verifies that the network interfaces are not in promiscuous mode. Upon detection, the file can be deleted from the system.

The primary aim of this project is to develop and demonstrate a fully functioning rootkit so as to find the vulnerabilities of the Linux operating system and try to exploit them. The exploits include key-sniffing, port hiding and unhiding, process hiding and unhiding, getting root access, and thrashing the operating system. This project also aims to detect exploits and prevent them from exploiting the system.

Related Work

Rootkits are a collection of software tools that are designed with the purpose of letting the attacker gain root-level access to a particular system and use that access for the attacker's personal gain, which may include thrashing the OS,exfiltration of data, key-sniffing, etc. Administrative privileges are required in order to install the rootkit. In this chapter, different methodologies for detection and prevention are explored. In this chapter, different methods of rootkit detection are compared for the rootkits that particularly affect the Windows operating system. The methods have been compared using various parameters, which include their relative strengths, weaknesses, performance, overhead, and ease of deployment.

This chapter specifically focuses on the detection of Linux kernel module rootkits through a statistical technique known as outlier analysis, which focuses on the comparison of the distribution of the system call addresses in a suspect system to that of a known uninfected system. This chapter introduces a new rootkit prevention mechanism called NICKLE. Nickle is a virtual machine monitoring-based system that uses a technique called memory shadowing, where the VM maintains a shadow memory for running the VM. All the guest kernel instructions are routed to the shadow memory after authentication; hence, only authenticated code is stored in the shadow memory and used to run the VM. This system is based on the fact that rootkits always try to execute their own kernel-level code.

This chapter focuses on developing a framework to classify different types of rootkits, which helps the admins determine the extent of the attacks and help in faster reactions to future attacks. Along with the development of the framework, the authors also tried to devise a mechanism to detect kernel-level rootkits. Once they establish the infection, they create new signatures in order to detect such rootkits in the future. The focus of this chapter is the prevention of persistent rootkits using a hypervisor-based file protection scheme. Since hypervisors have higher privileges than rootkits, this method uses that privilege to make critical system files read-only and unmodifiable by rootkits even with their kernel privileges. The authors also look to decrease the size of the hypervisor by combining it with the Bit Visor architecture.

This chapter talks about the history of the development of rootkits and the effects they have on a system. Prominent cases in the past involving root kits are discussed. Also, an overview of the methods to detect and eradicate rootkits is presented by the author(s). The objective of this chapter appears to be the analysis of rootkits and their effect on information security. This chapter reviews the ways in which Linux systems can be exploited, with a focus on kernel-based rootkits and countermeasures to tackle the rootkits. The efficiency of the different countermeasures is focused on rootkits that are classified by code injection and the flow of execution within the kernel.

This chapter focuses on detecting kernel-based rootkits through binary analysis. Binary analysis is executed to remove the kernel rootkit if the module's behaviour matches the behaviour of a rootkit. This technique is relied on even when attempts are made to hide the malicious modification in the kernel.

Proposed Architecture

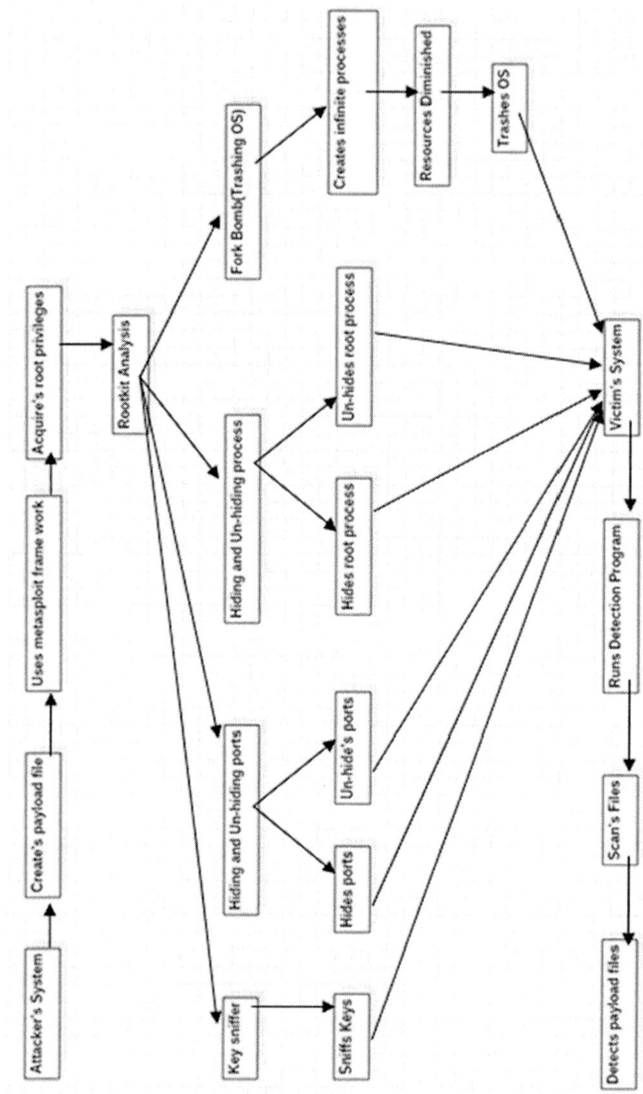

Figure 1: Rootkit Detection and Exploitation Architecture

- Creating a payload file: A payload in a metasploit refers to an exploit module. These exploits allow for a great deal of versatility and can be useful across numerous types of modules.
- Root Privileges: A payload file is exploited using exploit 'multi/handler'. For the payload to be exploited, the listening host IP address, and the listening port number are set on the attacker's system and then exploited. Using this, the access is obtained, all the modules of rootkit analysis can be inserted, and the necessary data can be extracted.

How to thrash an OS

Essentially, a function that calls itself twice every call and doesn't have any way to terminate itself is being created. It will keep doubling up until system resources are depleted, and thus the system has to be restarted to function normally.

Key Sniffer

Key Sniffer is a security vulnerability affecting some wireless keyboas. Wireless keyboards susceptible to KeySniffer allow attackers to eavesdrop on keystrokes.

Process of hiding and unhiding

This module controls who can access the information in the "/proc/[pid]' directories. The argument 'n' has three values: 0, 1, and 2. '0' indicates that everyone can access all '/proc/[pid]' directories. '1' indicates that users are denied access to files and subdirectories inside any '/proc/[pid]' directories excluding their own. '2' indicates that it works in a similar manner to mode 1, but in addition, the '/proc/[pid]' directories belonging to other users become invisible.

Port hiding

If malware is being used on a cloud server, the SSH connections, which are on port 22, would have to be hidden so that anonymity can be maintained. This method covers how to configure a firewall with various port-related rules.

Port un-hiding

If a cloud server is being used, incoming SSH connections on port 22 would have to be allowed in order to connect to and manage the server. This section covers how to configure firewalls with various port-related rules.

Detection programme (detection and prevention)

This programme is created to detect user and kernel-mode rootkits, specifically to locally check for signs of a rootkit. It is a shell script with some fragments written in C, trying to make it as platform-independent as possible since it runs on multiple Unix flavours. The rootkit, which gets root privileges, will first sniff the keys, hide and unhide UDP and TCP ports, and thrash the operating system. These modules then work towards hiding information from the user. For example, the key sniffer command shows a list of keys that are pressed by the user when the module is inserted. The most common technique for insertion of a rootkit in Linux is through a loadable kernel module, which uses multiple rootkit techniques. A loadable kernel module allows any information processed by the system to be modified. This runtime insertion of malicious code into the kernel can be deterred by turning off the ability to load kernel modules.

The rootkit first gains root access to the victim's system using the Metasploit framework, from which all modules are processed in the victim's system. For port hiding and unhiding, IP table commands are used to make and run a shell. For process hiding and unhiding, shell scripts are used. For thrashing the operating system, the fork bomb technique is used, in which a process is created and the process forks itself twice. This initiates four processes, and this goes on continuing and eats up all the resources of the system, thereby hanging the entire operating system and leaving no resources available to be taken by other processes. For key sniffers, code is developed in C and built using a make file, and a module is made to sniff keystrokes by the user. For rootkit detection, a detection programme that detects any interrupt evasion or the presence of any rootkit is used.

Modules and Implementation

- Creating a payload file

A payload in a metasploit refers to an exploit module. These exploits allow for a great deal of versatility and can be useful across numerous types of scenarios. Whether or not a payload is staged is represented by a "/' in the payload name. For example, "windows/shell_bind_tcp" is a payload with no stage that is widely used, whereas the payload file is used using the command "- msfvenom -p."

linux/x86/meterpreter/reverse_tcplhost=<ip address>lport =<listening port address> -f elf >name.elf" using the Metasploit framework. This payload file is installed on the victim's system.

- Root Privileges

Stagers set up a network connection between the attacker and victim and are designed to be small and reliable. It is difficult to always do both of these well, so the result is multiple similar *stagers*. Metasploit will use the exploit when it can. The payload file is exploited using the exploit 'multi/handler'. The payload to be exploited, the listening host IP address, and the listening port number are set on the attacker's system and then exploited. Using this, access is obtained, all the modules of rootkit analysis can be inserted, and the necessary data can be collected.

- Rootkit Analysis How to thrash an OS

In computing, a fork bomb is a denial-of-service attack wherein a process continually replicates itself to deplete available system resources, slowing down or crashing the system due to resource starvation.

:|: & ;:' is called a fork bomb. ' :()' means a function called ':' is being defined. ' :|: &' means run the function ':' and send its output to the ':' function again and run that in the background. The ';' is a command separator, like '&&'. ':' runs the function the first time. Essentially, a function that calls itself twice every call and doesn't have any way to terminate itself is created. It will keep doubling up until system resources are depleted. There is no prevention mechanism for such scripts, as they can run at the user level and damage the system.

Key Sniffer

Keystroke logging, often referred to as keylogging or keyboard capturing, is the action of recording (logging) the *keys* struck on a keyboard, typically covertly, so that the person using the keyboard is unaware that their actions are being monitored. The data can then be retrieved by the person operating the logging programme. There is a double array containing all pairs of keyboard characters; press the shift key as well as the character button. Whenever any key is pressed, it is matched to the characters in that array, and the matched characters are logged inside a file. A Linux module is used to make the code run at all times. Root privilege is required to run this.

Process hiding and unhiding:

- mount -o remount, hidepid=2 /proc
- The proc filesystem supports the following mount options:
- hidepid=n (since Linux 3)
- This option controls who can access the information in the /proc/[pid] directories. The argument, n, is one of the following values:
- 0: Everybody may access all /proc/[pid]
- 1: Users may not access files and subdirectories inside any /proc/[pid] directories but their own (the /proc/[pid] directories themselves) remain visible.
- 2: As for mode 1, but in addition, the /proc/[pid] directories belonging to other users become visible.

Port Hiding

If malware is being used on a cloud server, the SSH connections on port 22 will have to be hidden so one can be anonymous. This method covers how to configure a firewall with various port-related rules.

To hide all incoming SSH connections, the following commands are run:

- sudo iptables -A INPUT -p tcp --dport 22 -m conntrack --ctstateNEW,ESTABLISHED -j DROP
- sudo iptables -A OUTPUT -p tcp --sport 22 -m conntrack --ctstate ESTABLISHED -j DROP Preventing malware from getting root access is only the first step.

Port un-hiding

If a cloud server is being used, incoming SSH connections on port 22 will have to be allowed so that the user can connect to and manage the server. This section covers how to configure a firewall with various port-related rules.

To allow all incoming SSH connections, the following commands are run:

- sudo iptables -A INPUT -p tcp --dport 22 -m conntrack --ctstateNEW,ESTABLISHED -j ACCEPT
- sudo iptables -A OUTPUT -p tcp --sport 22 -m conntrack --ctstate ESTABLISHED -j ACCEPT Preventing malware from getting root access is only the first step.

Detection programme (detection and prevention):

This programme is created to detect user and kernel-mode rootkits, specifically to locally check for signs of a rootkit. It is a shell script with some fragments written in C, trying to make it as platform-independent as possible since it runs on multiple Unix flavours.

Searching for rootkit configuration files

In order to make rootkits flexible and customizable, the following steps are carried out:

- They are programmed to obtain their configuration from some system files, like "/dev/xxx.elf (payload files)" in the default configuration.
- The system files are scanned, and the payload files that can exploit the system are extracted.
- These files and their location in the system are shown in the report, which can be deleted, thus preventing the system from being exploited.

Results

Creating a payload file and getting root privileges

A payload is created and installed in the user system to get root privileges on the attacker's system.

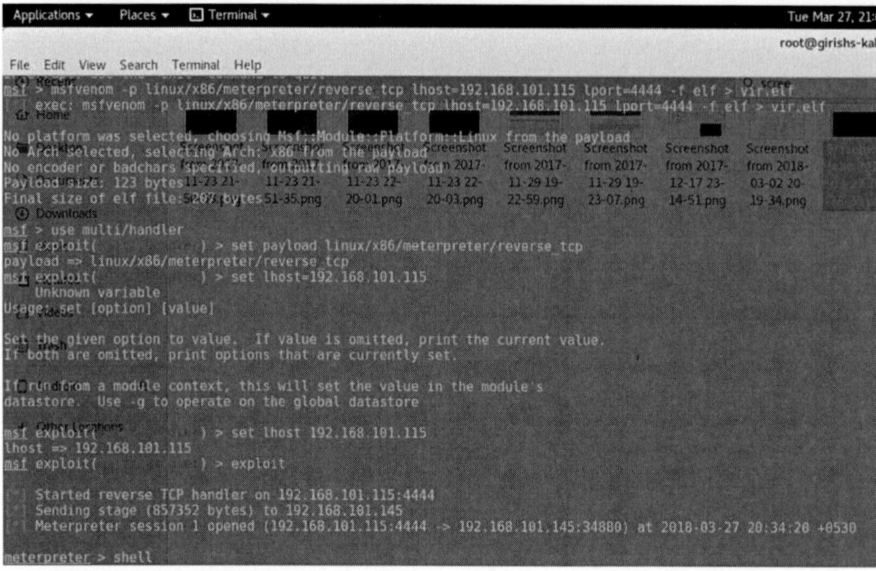

Figure 2: Creating payload vir.elf.

Figure 2 represents the creation of a payload using the Metasploit framework with the payload name "vir.elf," which is being exploited using a multi-handler that can be connected to the victim's system and can be exploited.

Gaining Root Access

Figure 3: Gaining root access

Figure 3 shows that the system is compromised and root privileges are obtained. The attacker's system is in the shell of the victim's system.

Rootkit Analysis: Key Sniffer

Key Sniffer is a Linux kernel module that is used to grab keys that are pressed on the keyboard, or a keylogger. Key Sniffer was originally written with the US keyboard (and conforming laptops) in mind. By default, it shows human-readable strings for the keys pressed.

The keypress logs are recorded in debug as long as the module is loaded. Only root can read the log. The module name has been camouflaged to blend in with other kernel modules. However, the script can be executed at shutdown or reboot to save the keys to a file.

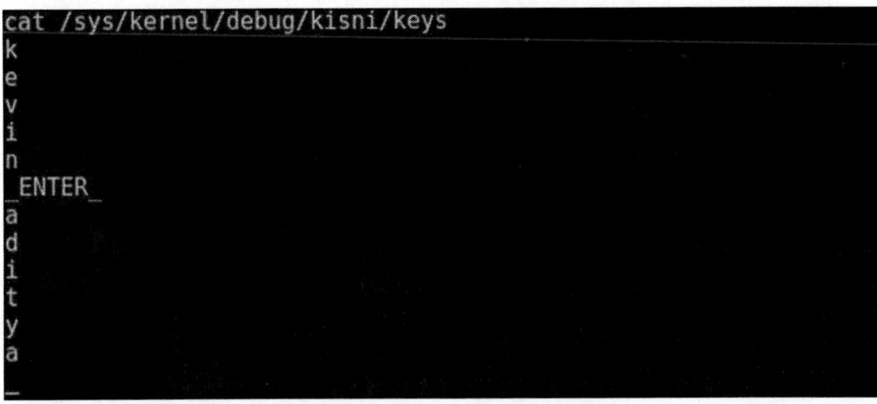

Figure 4: Keys sniffed in the attacker's system

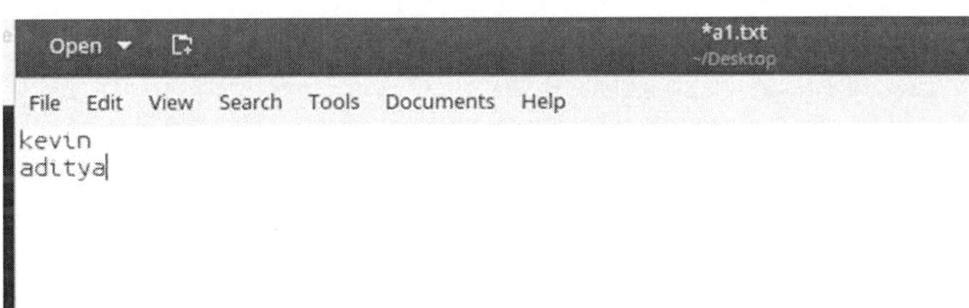

Figure 5: Keys pressed in the user's system

Figures 4 and 5 represent that the attacker has inserted the Key Sniffer module. The attacker's system prints the file of the keys pressed by the user in his or her system, and this module can be removed using RM Mod.

Hiding and unhiding processes

Root processes can be viewed by the user, but by making hide pid=2, the user's viewable processes are limited to only his processes and not root processes, to hide the malware activity as root processes, which can be unhidden by making hide pid=0.

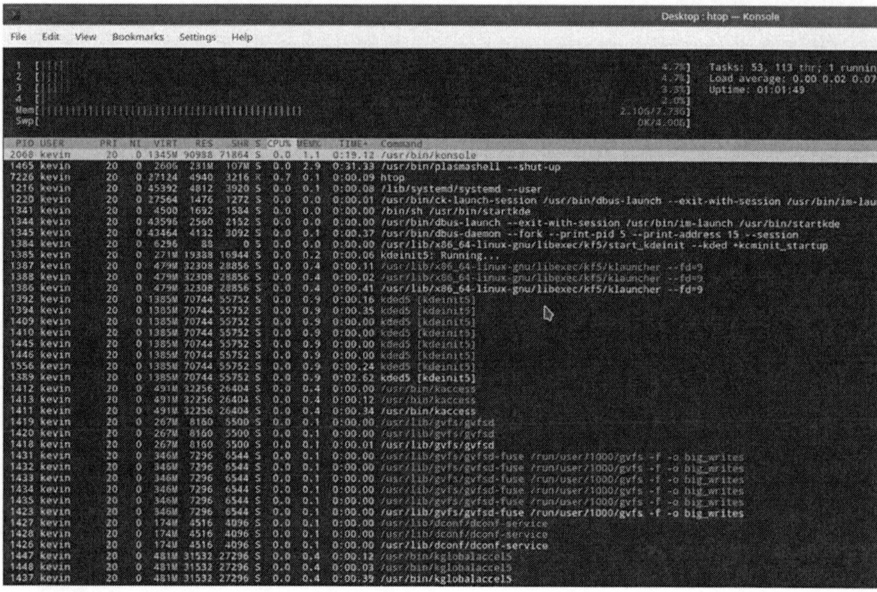

Figure 6: Hiding the root process

Figure 6 represents that the pid is set to 2 and the root processes are hidden, so the processes running in root cannot be identified by the victim.

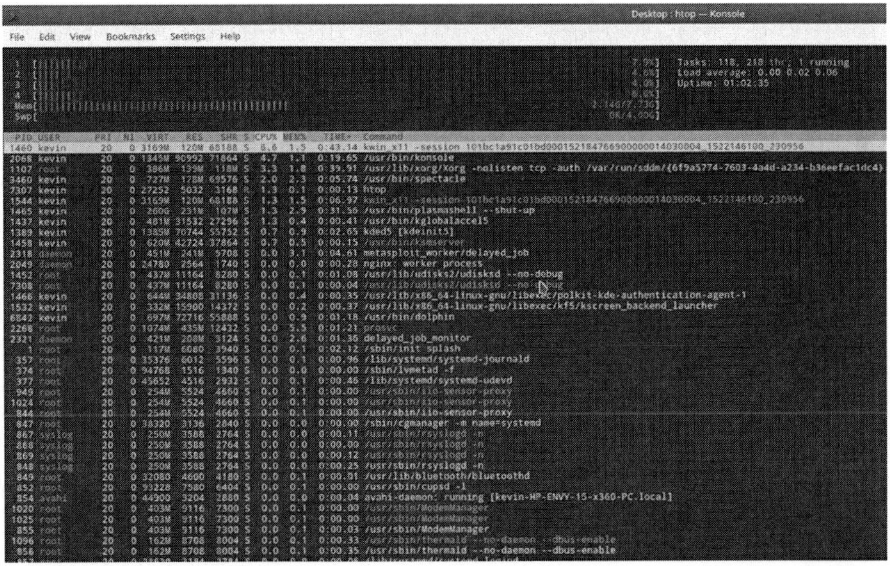

Figure 7: Un-hiding the root process

Figure 7 represents that the pid is set to 0 after the necessary operations are done by the attacker to unhide the root processes, and then the root processes can be viewed by the victim again.

Fork Bomb (throwing OS)

```
#!/bin/bash

:(){ :|: & };:
```

A PC restart is required because all the system resources are consumed by the code.

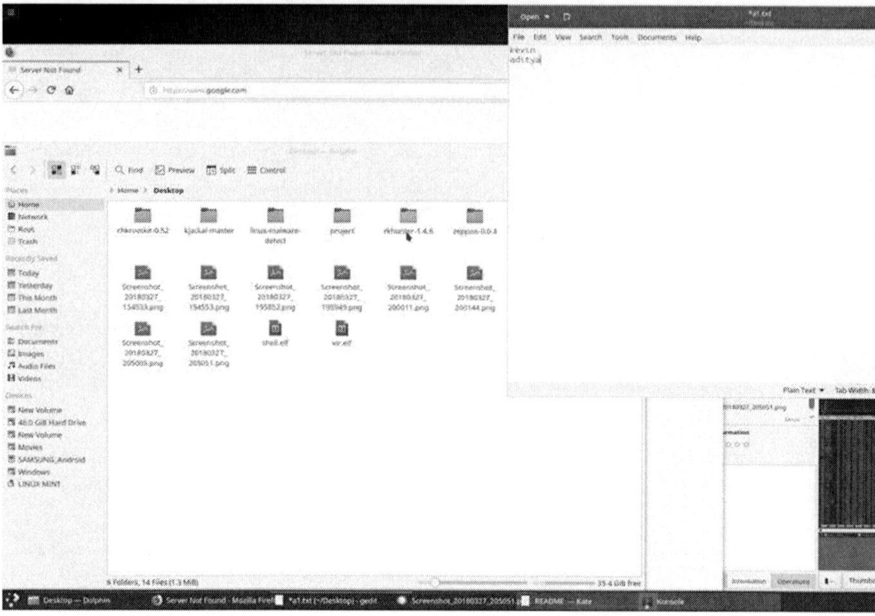

Figure 8: Thrashing OS (user system)

Figure 8 shows that the system resources of the victim are depleted thus leading to thrashing of the operating system and the victim has to restart to use the system again.

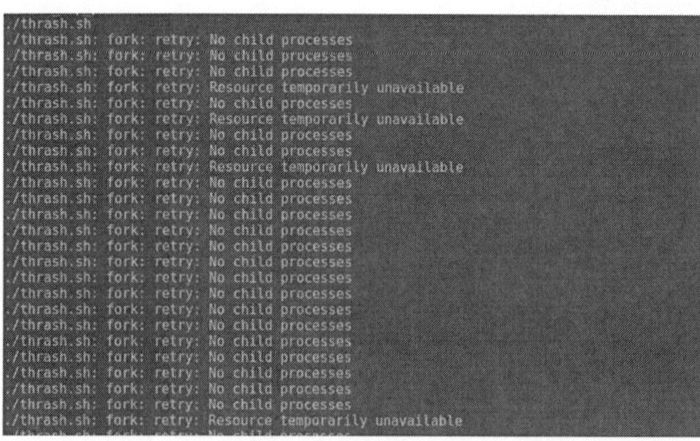

Figure 9: Thrashing OS (attacker's system) Figure 9 shows the resources of the victim being depleted in the attacker's system.

Hiding and Uh-hiding TCP Ports

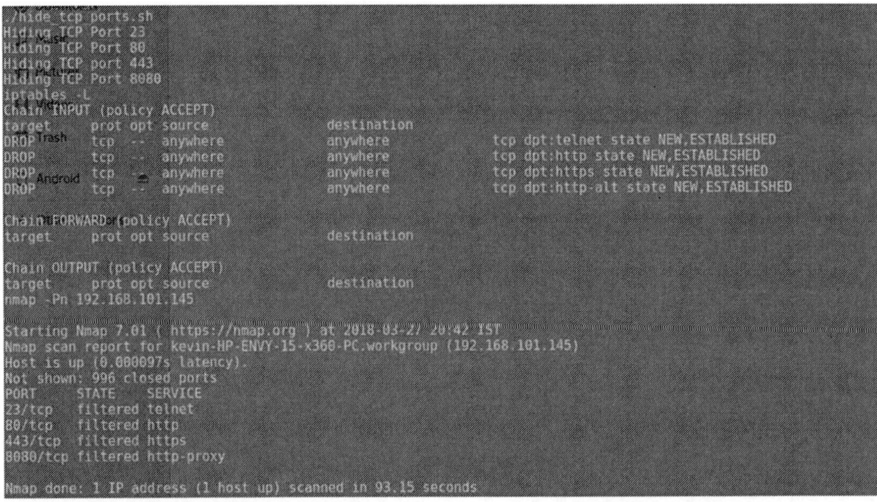

Figure 10: Hiding ports

Figure 10 shows that the ports are hidden using the drop command and then cross-checked using the iptables -L command and Nmap.

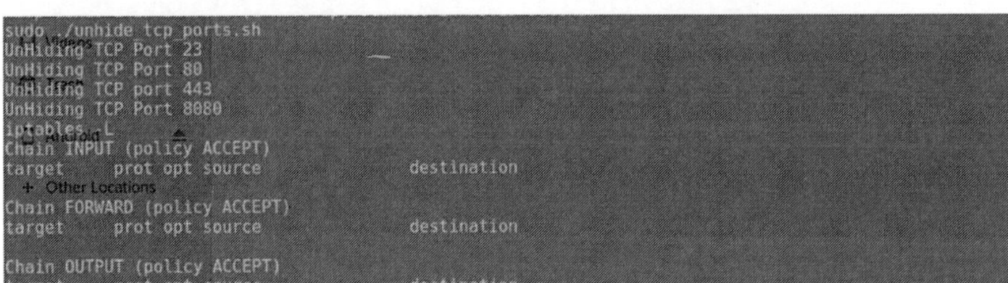

Figure 11: Un-hiding ports

Figure 11 shows that the ports are unhidden by using the accept command and using ip tables–Lit can be rechecked again.

DetectionProgram

This tool is created to prevent user and kernel-mode rootkits, specifically "to locally check for signs of a rootkit". This tool can be used to check all directories and show the rootkit file location in the report.

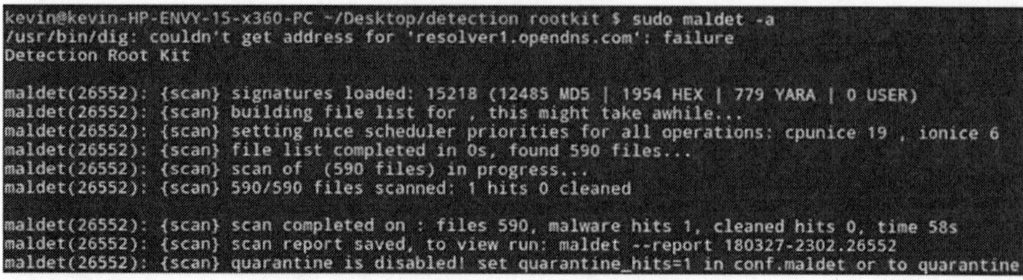

Figure 12: Detecting program files in the system

Figure 12 shows that the files are being loaded and the system files are scanned for detecting presence of payload file sand after the scan report is generated the location of detected payload files can be viewed.

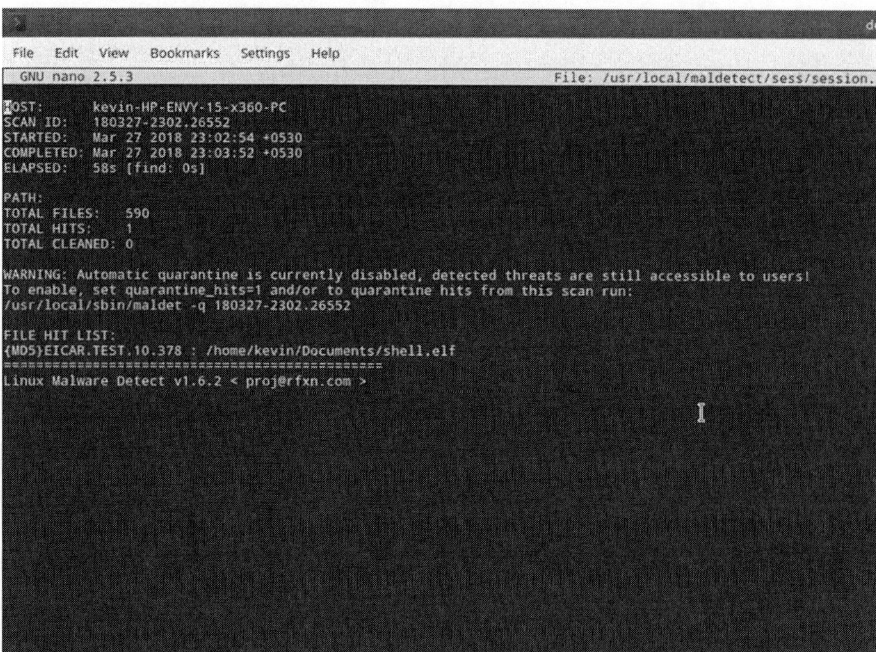

Figure 13: Scan report

Figure 13 represents the report of the scan and the file path of the detected payload file, which can be deleted.

Conclusion

Studying rootkits is very important because they will be with us for a while. Since there is no simple way to legislate against rootkits, an effective legal solution to the problem cannot be expected anytime soon. Moreover, rootkits are a very effective tool for hiding malicious software, and hackers will continue to use them as long as there is profit to be made. The only way they will be defeated is by the technology vendors making some fundamental shifts in the way they approach networked computing. Reduced permission levels, general bug fixes, and safe zones within a computer's memory space to install suspicious software are just a few of the techniques that can be used. But each new security technique has an accompanying complexity, which means that there will be flaws in the security software itself that can be exploited. The machine should always be updated by us. So, while technology may be the only way out of the problem of rootkits, it will be a long, difficult slog before a permanent solution is found. In conclusion, it is a good practise to secure the system from intruders, but it is important to know the effects of the intrusion. Protecting systems requires more than a single tool. It also requires pre-planning when installing new systems, maintaining backups, keeping up-to-date with the latest system and/or security patches, and performing regular audits. As long as there is a possibility of system vulnerabilities, security will continue to challenge IT professionals. Future work would involve developing hardware-based security for our detection system. Rigid security can be achieved by building hardware primitives that complement our system by providing better memory guarantees for both the kernel code and user-level applications.

References

1. Joy J, John A. Joy, James. 2011. "Rootkit detection mechanism: A survey.", Communications in Computer and Information Science book series (CCIS, volume 203).
2. Arnold, T.A. 2011. "A comparative analysis of rootkit detection techniques."
3. Wampler, D, Graham, J. 2011. "A method for detecting linux kernel module rootkits"
4. Riley, R. Jiang, X. Xu, D. 2008. "Guest-Transparent Prevention of Kernel Rootkits with VMM-Based Memory Shadowing". In: Lippmann R., Kirda E., Trachtenberg A. (eds) Recent Advances in Intrusion Detection. RAID 2008. Lecture Notes in Computer Science,

vol 5230. Springer, Berlin, Heidelberg. https://doi.org/10.1007/978-3-540-87403-4_1".
5. Levine, J. Grizzard, B and Owen H. L. 2006. "Detecting and categorizing kernel-level rootkits to aid future detection," in IEEE Security & Privacy, vol. 4, no. 1, pp. 24-32, doi: 10.1109/MSP.2006.11.
6. Yosuke, Chubachi. Takahiro, Shinagawa and Kazuhiko, Kato. 2010. "Hypervisor-based prevention of persistent rootkits". In Proceedings of the 2010 ACM Symposium on Applied Computing (SAC '10). Association for Computing Machinery, New York, NY, USA, 214–220. DOI: https://doi.org/10.1145/1774088.1774131"
7. Lynn Erla, Beegle. 2007. "Rootkits and Their Effects on Information Security", Information Systems Security, 16:3, 164-176, DOI: 10.1080/10658980701402049
8. Bunten, A. 2004." Unix and linux based rootkits techniques and countermeasures". In 16th Annual First Conference on Computer Security Incident Handling, Budapest.
[9] Kruegel, C. Robertson, W. & Vigna, G. 2004. "Detecting kernel-level rootkits through binary analysis", In *20th Annual Computer Security Applications Conference* (pp. 91-100). IEEE.

Reference Sites

1. [Ditt2002] Dittrich, D. 2002. "Root Kits" and hiding files/directories/ processes after a break-in. http://staff.washington.edu/dittrich/misc/faqs/rootkits.faq
2. [Fusy2002] "Fusys", Homepage kstat. http://www.s0ftpj.org/en/tools.html, 2002.
3. [Half1997] Halflife. 1997. "Abuse of the Linux Kernel for Fun and Profit" Phrack Magazine, Issue 50, Vol.7, File 5. http://www.phrack.org/phrack/50/P50-05.
4. [Hogl1999] Hoglund, G. 1999. "A real NT Rootkit, patching the NT Kernel" Phrack Magazine, Issue 55, Vol. 9, File 5. http://www.phrack.org/phrack/55/P55-06.
5. Boelen, M. Dudzinski, S. 2004. "Homepage Rootkit Hunter", http://www.rootkit. nl/projects/rootkit_hunter.html

Books

6. Lehti, R. Virolainen, P. 2004. "Homepage Aide", http://www.cs.tut.fi/~rammer/ aide.html

7. [BC2002] Bovet, D. Cesati, M. 2002. "Understanding the Linux Kernel", 2nd Edition. O'Reilly. ISBN- 0596002130.
8. [BTA1989] "Black Tie Affair", Hiding Out Under Unix. Phrack Magazin, Issue 25, Vol. 3, File http://www.phrack.org/phrack/25/P25-06.
9. [Bach1986] Bach, M. 1986. "The Design of the UNIX Operating System", Prentice-Hall. ISBN0132017997.
10. Prasad, V., Cohen, W., Eigler, F. C., Hunt, M., Keniston, J., and Chen, B. Locating System Problems Using Dynamic Instrumentation. In OLS '05.
11. Riley, R., Jiang, X., and Xu, D. Guest-Transparent Prevention of Kernel Rootkits with VMM-based Memory Shadowing. In RAID (2008).
12. Forrest, S., Hofmeyr, S. A., Somayaji, A., and Longstaff, T. A. A Sense of Self for Unix Processes. sp 00 (1996), 0120.
13. Yin, H., Liang, Z., and Song, D. HookFinder: Identifying and Understanding Malware Hooking Behaviors. In NDSS (2008).
14. Nick L. Petroni, J., and Hicks, M. Automated Detection of Persistent Kernel Control-flow Attacks. In CCS '07: Proceedings of the 14th ACM conference on Computer and communications security (New York, NY, USA, 2007), ACM, pp. 103–115.

CHAPTER 9

ENHANCING AI APPLICATIONS USING BLOCKCHAIN

SUGANESHWARI G[1], SUDHA A[2]

[1]SCHOOL OF COMPUTER SCIENCE AND ENGINEERING, VELLORE INSTITUTE OF TECHNOLOGY, CHENNAI

[2]CENTER FOR SMART GRID, SCHOOL OF COMPUTER SCIENCE AND ENGINEERING, VELLORE INSTITUTE OF TECHNOLOGY, CHENNAI

Abstract

Blockchain is a decentralized, immutable database that allows many parties to exchange encrypted data instantly, transparently, and concurrently as they start and complete transactions. Permissioned members gain trust and confidence in their interactions with other firms since they share the same truth. Artificial intelligencc imitates how the human mind makes decisions and solves problems. AI's advantages are automating monotonous activities, better decision-making, and improved customer experiences. Moreover, automation and augmentation in AI applications utilize sensitive data, which can be tampered with due to a lack of authentication. Authenticity is the key component of blockchain that can bring authenticity to AI. With blockchains, especially smart contracts, AI can improve blockchain protection in big data. The industries that would improve with the combination of AI and blockchains are finance, health care, and military sectors. In this chapter, we will discuss in detail how blockchains and AI can be combined, as blockchains bring security or trust to the data, and AI can utilize the trusted data.

Keywords: Blockchain, Artificial Intelligence, Trust, Data Security, Automation

Introduction to Blockchain

A data structure that promotes decentralization, security, and transparency while preserving transactional records is known as a blockchain. Blockchains are distributed ledgers that are entirely available to everyone on the network. Once placed on a blockchain, data cannot be changed or edited easily. The original files can be found in the local recording office, which may eliminate the need to scan the paperwork. Owners may trust that their deed is correct and permanently recorded if property ownership data is kept and verified on the blockchain. Blockchains are generally divided into permissioned, permissionless, or both, as shown in Figure 1. Among these broad categories, blockchains are further classified as public or hybrid. The public is entirely permissionless, a hybrid might be permissioned or permissionless, and consortium and group-based are permissioned categories.

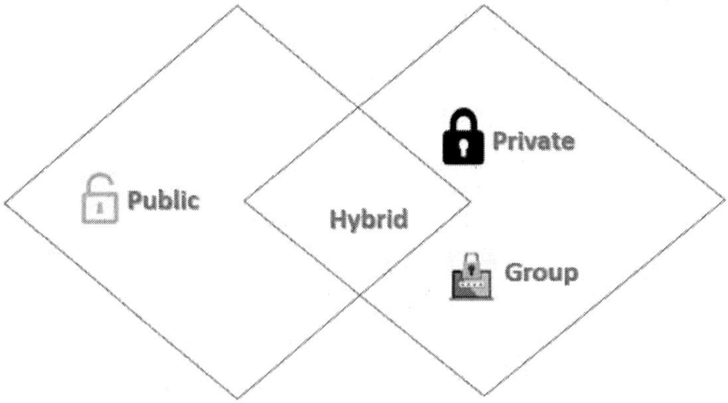

Figure 1: Types of Blockchain

Considering the broad categories, the blockchain model is further classified into four types as follows:

Public Blockchains

Public blockchains are permissionless, decentralized, and accessible to anyone. All nodes have equal access to the blockchain in public blockchains and can upload and validate new data blocks. The main uses of public blockchains are for mining and trading cryptocurrencies. Popular public blockchains include Litecoin, Ethereum, and Bitcoin. On these open blockchains, nodes generate blocks for network transactions by resolving cryptographic equations to mine cryptocurrency. Miner nodes are compensated with a small amount of cryptocurrency in exchange for their efforts. The miners essentially work by formulating a transaction and receiving a fee in exchange for their mining efforts.

Private Blockchains

Permissioned blockchains that a single organization manages are called private or managed blockchains. A central organization selects the nodes for a private blockchain. Additionally, the central authority occasionally gives every node the same task permission. Private blockchains are only partially decentralized because they have little public access. Transients, a business-to-business trading platform for virtual currencies, and Hyperledger are examples of private blockchains.

Consortium Blockchains

Federated blockchains, also known as consortium blockchains, are permissioned blockchains that several different organizations manage. For greater security, group-based blockchains are more decentralized than private ones. Furthermore, group formation can be challenging because it calls for cooperation among numerous organizations, posing logistical difficulties and risks.

Hybrid Blockchains

A single organization manages hybrid blockchains, but they are subject to rules set forth by the public blockchain, which are necessary to perform particular transaction validations. A hybrid blockchain, like IBM Food Trust, was developed to boost productivity throughout the food supply chain (Wegrzyn, 2021, 2).

Blockchain benefits, Platforms, and Uses

Blockchain platforms have complex, powerful programming languages that manage various Web3 functions, transaction initiation and execution, and smart contract creation.

Ethereum

One well-known blockchain platform is Ethereum (ETH). Users can use it to develop brand-new financial applications, decentralize markets, develop games, make bitcoin wallets, and do much more. This platform's main objective is to stop third parties from accessing data stored for future financial tracking. Most core protocol developers, crypto-economic researchers, cypherpunks, and miners are on Ethereum. The features of the ETH include Smart Contracts, tokenization, and support for C++ and Python programming languages (Arroyo, 2021, 2).

Bitcoin

A peer-to-peer electronic cash system is used instead of working through a banking institution to make direct and uninterrupted payments from one to another. Also, digital signatures will support the problem, but key benefits are lost if a trustworthy third party needs to prevent double-spending. To establish timestamps, the network hashes each transaction into a line of hash-based proof-of-work. This creates a record that can only be altered by reiterating the proof of work. The lengthiest chain shows the sequence of the observed events and their origin from the most CPU power. If nodes are not working together to attack the network and take control of most of the CPU power, they will create the longest chain and outperform attackers. The network itself only needs a little structure. Nodes can leave and rejoin the network whenever they like, using the longest proof-of-work chain to verify what happened during their absence (Nakamoto, 2008, 2). The finest efforts are used to broadcast messages.

Avalanche

Avalanche, which claims itself as the quickest smart contracts platform on the market, enables users to establish personalized private and public blockchains, scale quickly with little infrastructure, and develop decentralized apps quickly and affordably. Additionally, it uses CPU-Optimal energy efficiency, which keeps machines reasonably quiet when utilizing the platform. The correct nodes are led to the same outcome by the avalanche mechanism, which repeatedly collects random samples from the network. The key idea is derived from gossip algorithms. In the Avalanche, each node periodically queries k randomly selected other nodes while captivating the result returned by most of the questioned nodes (Tanana, 2021, 1-3).

Ripple

In addition to developing its cryptocurrency, XRP, Ripple also developed a digital payment network and protocol based on blockchain technology. A payment settlement, asset exchange, and remittance system make up the Ripple system's core. Instead of employing blockchain mining, Ripple uses a consensus process using a network of computers owned by banks to confirm transactions. While bitcoin transactions require more energy, take longer to confirm, and have higher transaction fees, ripple transactions use less energy, ensure more quickly, and have much lower transaction costs (Hijfte, 2020, 239).

Hyperledger

Developing a reliable framework, tools, and libraries for permissioned, enterprise-grade blockchain installations is the focus of the Hyperledger open-source community. It is an international partnership that the Linux Foundation is hosting and has member organizations from the financial, banking, IoT, supply chain, manufacturing, and technology sectors. There are numerous sub-projects, such as Hyperledger Cello, Hyperledger Fabric, Hyperledger Sawtooth, Hyperledger Composer, Hyperledger Explorer, and Hyperledger Caliber.

Other than these platforms, many blockchain platforms work for various cryptocurrencies, such as IBM Blockchain, Cardano, etc. (Tanana, 2021, 1-3).

AI Applications Integration with Blockchain

Blockchain's digital ledger provides information about the theoretical foundations of AI and the provenance of the data it consumes and provides a solution to the dilemma of explainable AI. This increases trust in the data's accuracy and, as a result, in the AI recommendations. Blockchain-based storage and distribution of AI models result in the creation of an audit trail, which boosts data security. Blockchain and AI have the potential to provide new value to multi-party company operations by lowering barriers to entry, boosting efficiency, and quickening transaction times. For instance, AI models incorporated into blockchain-based smart contracts can recommend recalling outdated goods. The resolution of disputes and choice of the most ecologically friendly shipping method is determined by transactions like stock purchases, reorders, and payments based on predetermined criteria and events.

Blockchain Integration with Healthcare Domain

Our healthcare system satisfies our health demands, and as our population ages, productivity expectations rise even higher (Hollander, 2007, 34-45). Additionally, the value of real-time data is evident, as in the instance of COVID-19. With COVID-19, information on a patient's health (such as temperature or symptoms) is beneficial. The aging population and the prevalence of chronic diseases have increased health awareness and the demand for better patient care. Now, the trend is for consumers to prefer patient-centered healthcare to traditional healthcare. Mobile healthcare, which has recently evolved into universal healthcare, has gained popularity at the expense of patient- and hospital-centered healthcare. Assuring the security of every piece of equipment that can connect to the internet is the biggest challenge facing the electronic and ubiquitous healthcare sector (Islam, 2020, 106627). Many networks today are built to function with mobile devices and offer mobile-focused client services.

A remarkable amount of medical data is produced and shared every day, every week, due to the growth of Internet-based medical devices and apps. Although there may be some benefits to the current healthcare system, there are also some drawbacks. Most of these drawbacks include financial, technological, standardized, individual attitudinal, and organizational restraints (Zheng, 2018, 352-375).

The contemporary decentralization mechanism of blockchain offers reliable answers to problems pertaining to the security and dependability of the healthcare system. The absence of intermediaries is another compelling feature of blockchain technology. Blockchain has recently been used to create new pathways and patterns outside cryptocurrency (Yaeger, 2019, 1-7). Without intermediaries, this would be a novel delivery method that might pave the way for the development of health care. It integrates blockchain technology into healthcare applications, considering all privacy-related issues, security reliability, and patient and electronic health record access. Blockchain technology is rapidly growing in the healthcare sector in the digital age. Potential applications of blockchain technology include cryptocurrency (Nakamoto, 2008, 2), commerce, banking, advertising, insurance, and copyright protection.

Microtransactions involving the production and transfer of smart contracts, decentralized exchange, digital assets, and consensus mechanisms are some of the potential currently provided by technology. In the current environment, distinct blockchain application uses smart contracts (Szabo, 1996, 28). Time and money savings are made possible via intelligent contracts. One of the most popular and recommended frameworks for building an intelligent contract is Ethereum. Blockchain-based intelligent contracts have established a new paradigm across various businesses. Smart contracts are used in the healthcare industry to strengthen the conventional method greatly. Smart contracts are more straightforward, and solutions to optimizing people's workflow are quicker. They ensure security while still being simple to use when sharing sensitive information. The lack of intermediary parties is another compelling feature of a smart contract. Blockchain has recently created new trends and opportunities outside of cryptocurrencies.

The ability of the healthcare sector to adapt to the blockchain has revolutionized (Hölbl, 2018, 470). The health ecosystem prioritizes patients, and one of its main objectives is to increase health data security, privacy, and interoperability. The health sector generates enormous amounts of information, including patient medical records, data on clinical trials, billing, medical research, and many others (Bach, 2018, 1545–1550). The popularity and use of blockchain technology are changing throughout time. As a result, blockchain technology in healthcare has become a significant problem for all fields and researchers, as shown in Figure 2.

Distributed ledgers, the Hyperledger framework, a mechanism proposed as smart contracts and proof-of-concept, are the main components of the blockchain application field for this type of contribution. The application fields include the most recent blockchain frameworks, like Hyperledger and Ethereum. To construct an application in healthcare, this proposes many design considerations linked to private chains, Proof-of-Concept (PoC), Proof of Stake, and Proof-of-Work. PoC, private blockchain, and smart contract-based solution creation over Ethereum and the Hyperledger architecture are the key application areas for this section.

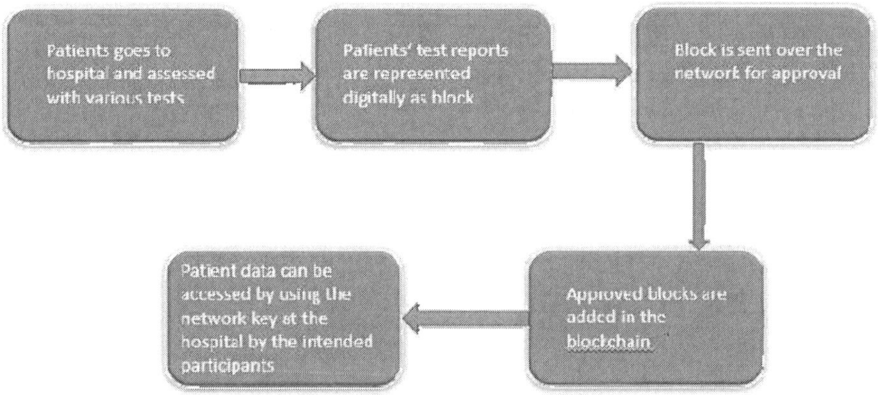

Figure 2: Blockchain in Health care

Blockchain and IoT

The Internet of Things, or IoT, has generated a lot of research attention because of its varied range of applications and its potential for energy efficiency, enhancing flexibility, and easy implementation that strong integration can provide. Blockchain technology's rapid advancement makes this future more accessible than ever. However, IoT users' privacy could be jeopardized due to the requirement for remote analysis data collecting and cloud integration. Over time, the meaning of IoT has grown to include consumer and household electronics and industrial equipment. The Industrial Internet of Things (IIoT) is the standard term to describe this (IoE).

Since the beginning of the IoT, security has been a concern. Significant research has been done describing the possible security risks of IoT due to the mix of low-powered equipment, low cost, challenges with fixes and upgrades, and wireless networking. Blockchain technology has been applied to developing cryptographic protocols or improving security in IoT contexts and ecosystems. Privacy-preserving Thin-client Authentication Scheme (PTAS) (Gupta, 2020, 47–56) to improve security for users of IoT devices with limited storage (PIR). Additionally, thin-client data is shielded from collaboration with the (m1) private PTAS by no more than (m1) full-node users. Evaluating their computational and communication overhead demonstrates these two techniques' potential. A blockchain-based system that uses the Physically Unclonable Function (PUF) architecture to improve authentication while maintaining user privacy was recently presented (Dilawar, 2020, 82–89). Accordingly, comparing blockchain smart contracts, cryptographic methods, and trusted third parties (Dubovitskaya, 2017, 650), none of these methods alone can guarantee privacy preservation. To securely share geolocation information in a trusted execution environment, they put into practice a hybrid strategy that integrates the methodologies above. In order to maintain users' privacy without lowering the quality of the service, Khezr (2022, 1736) developed a location privacy preserving strategy with a similar goal of focusing on privacy when employing location-based services. Although it is currently at the proof-of-concept stage, it provides exceptional service.

Freight transportation, log operational maintenance data, and component tracking and compliance are some IoT and blockchains applications. Moving freight is difficult and involves numerous stakeholders with various priorities. A blockchain with IoT support may monitor shipping container temperatures, locations, arrival times, and status. All stakeholders may put their trust in the data and move quickly to transfer products thanks to immutable blockchain transactions. IoT devices maintain track of the maintenance and security status of essential machines. By building a tamper-proof database of operational data, blockchain enables the support of whole things from engines to cranes. Partners for third-party maintenance can keep an eye on the blockchain for preventive maintenance and register their work back in. Governmental bodies may also be given access to operational documentation to confirm compliance. For safety and regulatory compliance, tracking the components that go into a vehicle, aircraft, or another item is essential. The provenance of components can be tracked by all parties throughout a product's life thanks to IoT data that is kept in open blockchain ledgers. Sharing this information with monitoring bodies, shippers, and manufacturers is risk-free, simple, and economical.

Blockchain and Transportation

Some of the IoT's promises, such as improved communications to support networks and quick reporting on events like accidents or road conditions, can be realized with the rise of connected and autonomous vehicles and vehicular networks. Such information comprises vehicle identity information, vehicle movements, messages sent over networks, and residual locations, which poses possible privacy hazards. Numerous studies have been conducted on integrating blockchain technology and privacy protection in vehicular ad hoc networks (VANETs). BlockAPP was created by Sharma and Chakraborty (2018, 1-6) to authenticate automobiles in the Internet of Vehicles while maintaining their privacy (IoV). A privacy-trust paradigm for VANETs (Lu, 2018) and uses blockchain technology to record certificate and revocation transparency. The authors introduced a privacy-preserving authentication method (Lu, 2019) in an expanded blockchain setting for VANETs. Guehguih et al. (2019, 16-21) proposed a dual blockchain system in which private blockchains were used for authentication and public blockchain were used to manage event messages as a means of privacy-preserving authentication in VANETs. Carpooling is a problem where users' private information is at risk due to vehicular fog computing (Li, 2017). The authors created a privacy-preserving authentication method with blockchains to verify the messages exchanged in VANETs (Feng, 2020).

Li et al. created CreditCoin, a privacy-preserving blockchain-based announcement network with incentives (VSNs), to address the issue of user anonymization and a need for more motivation to forward announcements in Vehicular Social Networks (Li, 2019, 4573-4584). CreditCoin supports the anonymous vehicular announcement aggregation protocol for sending announcements. Despite its anonymity, the technique enables the identification of malicious users. A method for concealing a vehicle's identification in a VSN (Pu, 2020, 308-324). The VSN's data is kept on a blockchain to prevent manipulation, and the consensus is achieved via Practical Byzantine Fault Tolerance (PBFT). In Shen et al. (2020, 6610), the authors recommended integrating a consortium blockchain with a location-based service that ensures user privacy to overcome the man-in-the-middle attack problem and the requirement for certificate authority while interacting with resource-constrained smart vehicles. With PPVC, a tiny certificate authority, the location-based service protocol that protects privacy was created. The experimental evaluation and security analysis demonstrate the effectiveness of this strategy in a VSN environment.

According to Baza (2020, 1-6), their suggested public blockchain for ridesharing could resolve several challenges caused by traditional central third-party ridesharing techniques, including privacy issues. The drivers can conduct ridesharing services without requiring assistance from a centralized source. Sensitive information like locations, dates, and times are protected on the client's end by sending cloaked queries, to which the drivers then provide encrypted offers in response. This approach, which has shown viability in actual use, was implemented using the Ethereum smart contract platform. Along with working on privacy-enabled ridesharing, the goal was to solve the problem of service providers sharing data across organizational boundaries. To do this, they developed *CoRide*, a hailing service with privacy protection that logs rides using fog computing and a consortium blockchain (Li and Zhu 2019, 4573-4584). Passengers and drivers are matched using smart contracts; this process was also developed utilizing an Ethereum network. However, as ridesharing platforms have their technology stacks and formats, it is expected in this study that all users would utilize the same data formats for payment and matching, which is rarely the case.

Emerging Trends

Integration with Data Analysis and Artificial Intelligence

Making decisions about patient care, medical research, and clinical trials may need to be reconsidered with the help of artificial intelligence tools and blockchain technology. Deep learning and machine learning are used to achieve this (Ali, 2020, 208-222). Blockchain and artificial intelligence work together to improve decision-making in the medical domain. The study is significant because the prediction's validity and accuracy are still unknown. The justifications for medical records and diagnoses are also hazy. Finding links between two blockchain-based documents is among the benefits of blockchain data analytics and aggregation. It eliminates the requirement for processing services and additional storage, enhancing report legitimacy and avoiding report skewing (Mamoshina, 2018, 5665).

Integration of Blockchain with Cloud Computing

The amount of data generated and held in each node is one of the significant issues with a blockchain. With more nodes and transactions, this issue will only worsen. So, reducing energy consumption while increasing capabilities is one feasible prospective solution. Researchers have investigated the usefulness of cloud computing in healthcare settings. In order to provide stable, affordable heterogeneous healthcare settings in IoT networks, some studies in the healthcare field have developed more reliable and efficient data management systems using a distributed cloud architecture that combines blockchain technology and cloud computing (Ali, 2021, 23-43). The abundance of data and the speed of decision-making will improve medical decisions. We know that only a few operations have produced a blockchain-based framework designed for the healthcare industry. However, cloud computing still requires research in modeling, architecture and cost-benefit analysis, and other factors.

Privacy Preservation with Blockchain Technologies

Future work will involve many architectural characteristics that, when realized, will be highly advantageous. There is a need for more blockchain and smart contract architectures for various applications in an industry-neutral way. As a result, developers and researchers might apply best-of-breed variations with the knowledge that their application is appropriate for the task. The templates and reference architectures generated from this can help across various industries. This would not apply to all fields because different implementations might need different attributes.

Given the ongoing research and development in IoT, there is a rising need for work to integrate blockchain and smart contract paradigms into these developing ecosystems. Importantly, driven by success, the Internet of Things promises are closer than ever thanks to several concepts and additional consensus algorithms. More IoT ecosystems will consequently actively use blockchain and consensus solutions. The privacy of tools, systems, people, and data must be safeguarded. It is also anticipated that security procedures and methods will advance, offering more options for ensuring privacy. Preserving biometric templates in IoT environments is crucial because the IoT environment has been decentralized by blockchain, unlike databases, where they are stored in a centralized fashion. This is because biometric authentication technologies have made it easier to manage IoT access. As a result, the next exciting field for research should be decentralized biometric authentication systems with template protection mechanisms.

Use Case of Blockchain Technology: Smart Home

We have multiple IoT-sensitive devices connected at home, which transmit data to the smartphone, email, and computer via the internet. Since cloud storage is used to share and store data to comprehend security in terms of confidentiality, integrity, and availability, the smart home uses privacy preservation. The blockchain miners manage all the incoming and outgoing information to help the blockchain, and each transaction is stored in termed blocks. Thus, the hyper ledger platform tools can also be used with IoT devices (Dorri, 2017, 618-623).

Conclusion

The study demonstrates blockchain use and integration, particularly in healthcare, IoT, and transportation. Developing secure, more decentralized systems with exceptional data processing and management levels has another significant potential benefit of blockchain technology. This study aims to enlighten current advancements in blockchain technology-based applications. With a larger range, many blockchain potentials remain unrealized. Many research projects include novel architecture, platforms, and related models, but they are less likely to include blockchain components used to automate blockchain operations. Additionally, blockchain technology is still in its infancy and is mostly used to integrate and incorporate approaches in its early phases of development. However, it is used exponentially in situations considered rational and necessary.

References

1. Ali, Farman, Shaker El-Sappagh, SM Riazul Islam, Daehan Kwak, Amjad Ali, Muhammad Imran, and Kyung-Sup Kwak. 2020. "A smart healthcare monitoring system for heart disease prediction based on ensemble deep learning and feature fusion". Information Fusion 63. 208-222.
2. Ali, Farman, Shaker El-Sappagh, SM Riazul Islam, Amjad Ali, Muhammad Attique, Muhammad Imran, and Kyung-Sup Kwak. 2021. "An intelligent healthcare monitoring framework using wearable sensors and social networking data." *Future Generation Computer Systems*. 114 (2021): 23-43.
3. Arroyo J, 2021. Keynote Speech "An overview of Decentralized Autonomous Organizations in the blockchain, 2021". *Third International Conference on Blockchain Computing and Applications (BCCA)*, 2021, 2-2.
4. Bach L.M, B. Mihaljevic, and M. Zagar. 2018. "Comparative analysis of blockchain consensus algorithms." *Proc. 41st Int. Conv. Inf. Commun. Technol., Electron. Microelectron. (MIPRO),* May 2018, 1545–1550
5. Baza M, M. Mahmoud, G. Srivastava, W. Alasmary and M. Younis. 2020. "A light blockchain-powered privacy-preserving organization scheme for ride sharing services. " *Proc. IEEE 91st Veh. Technol. Conf. (VTC2020-Spring)*. 1-6.
6. Dilawar N, M. Rizwan, F. Ahmad, and S. Akram. 2019. "Blockchain: Securing Internet of medical things (IoMT). " *Int. Adv. Comput. Sci. Appl.*, 10, 1, 82–89, doi: 10.14569/ijacsa.2019.0100110.
7. Dorri A, S. S. Kanhere, R. Jurdak, P. Gauravaram. 2017. "Blockchain for IoT security and privacy: The case study of a smart home. "*In IEEE international conference on pervasive computing*

and communications workshops (PerCom workshops), 618-623, IEEE.

8. Dubovitskaya A, Z. Xu, S. Ryu, M. Schumacher, and F. Wang. 2017. Secure and trustable electronic medical records sharing using blockchain. in *Proc. AMIA Annu. Symp.*, 2017, p. 650.
9. Feng Q, D. He, S. Zeadally and K. Liang. 2020. Bpas: Blockchain-assisted privacy-preserving authentication system for vehicular ad hoc networks. *IEEE Trans. Ind. Informat.*, vol. 16, no. 6, pp. 4146-4155, Jun. 2020.
10. Guehguih B and H. Lu. 2019. "Blockchain-based privacy-preserving authentication and message dissemination scheme for vanet." *Proc. 5th Int. Conf. Syst. Control Commun.*, pp. 16-21.
11. Gupta S, V. Malhotra, and S. N. Singh, 2020. Securing IoT-driven remote healthcare data through blockchain," *in Advances in Data and Information Sciences. Singaporess:* Springer. 47–56.
12. Hollander M, Chappell N, Prince M, and Shapiro E. 2007. "Providing care and support for an aging population: Briefing notes on key policy issues." *Healthcare Quart.* 10(3), 34–45.
13. Hölbl M, M. Kompara, A. Kamišalić, and L. Nemec Zlatolas. 2018. "A systematic review of the use of blockchain in healthcare. " *Symmetry*, vol. 10, no. 10, p. 470, Oct. 2018
14. Islam, Anik, and Soo Young Shin. 2020. "A blockchain-based secure healthcare scheme with the assistance of unmanned aerial vehicle in Internet of Things." Computers & Electrical Engineering 84 (2020): 106627.
15. Khezr S, M. Moniruzzaman, A. Yassine, and R. Benlamri. 2019. "Blockchain technology in healthcare: A comprehensive review and directions for future research". *Appl. Sci.*, 9, 9, 1736, Apr. 2019, doi: 10.3390/app9091736.
16. Li M, L. Zhu and X. Lin. 2019. "Efficient and privacy-preserving carpooling using blockchain-assisted vehicular fog computing. *IEEE Internet Things J.*, 6, 3, 4573-4584.
17. Lu Z, W. Liu, Q. Wang, G. Qu and Z. Liu. 2018. "A privacy-preserving trust model based on blockchain for vanets", *IEEE Access*, 6, 45 655-45 664.
18. Lu Z, Q. Wang, G. Qu, H. Zhang and Z. Liu. 2019. "A blockchain-based privacy-preserving authentication scheme for vanets. " *IEEE Trans. Very Large Scale Integr. (VLSI) Syst.*, 27. 12, 2792-2801.
19. Mamoshina P, L. Ojomoko, Y. Yanovich, A. Ostrovski, A. Botezatu, P. Prikhodko, E. Izumchenko, A. Aliper, K. Romantsov, and A. Zhebrak. 2018. "Converging blockchain and next-generation

artificial intelligence technologies to decentralize and accelerate biomedical research and healthcare. " *Oncotarget*, 9, 5, 5665, 2018.
20. Nakamoto S. 2008. Bitcoin. "A peer-to-peer electronic cash system." Bitcoin.–URL: https://bitcoin. org/bitcoin. pdf 4, no. 2.
21. Pu Y, T. Xiang, C. Hu, A. Alrawais and H. Yan. 2020. "An efficient blockchain-based privacy preserving scheme for vehicular social networks. " *Inf. Sci.*, 540, 308-324.
22. Sharma S and S. Chakraborty. 2018. "Blockapp: Using blockchain for authentication and privacy preservation in iov." *Proc. IEEE Globecom Workshops*, 1-6.
23. Shen H, J. Zhou, Z. Cao, X. Dong and K.-K. R. Choo. 2020. "Blockchain-based lightweight certificate authority for efficient privacy-preserving location-based service in vehicular social networks. " *IEEE Internet Things J.*, 7, 7, 6610-6622.
24. Szabo, Nick. "Smart contracts: building blocks for digital markets." *EXTROPY: The Journal of Transhumanist Thought*, (16) 18, no. 2 (1996): 28.
25. Tanana, Dmitry. "Avalanche blockchain protocol for distributed computing security." *In 2019 IEEE International Black Sea Conference on Communications and Networking (BlackSeaCom).* IEEE,1-3, doi:10.1109/ BlackSeaCom. 2019. 881 2863.
26. Van Hijfte, Stijn. 2020. "Blockchain platforms: A look at the underbelly of distributed platforms." *Synthesis Lectures on Information Concepts, Retrieval, and Services* 8, no. 1 (2020): i-239.
27. Wegrzyn K, Wang E. 2021. "Types of blockchain: public, Private, or something in between. Folcy & Lardner LLP" 2021 Aug 19 [accessed 2022 May 31. https://www.foley.com/en/insights/publications/2021/08/types-of-blockchain-public-private-between.
28. Yaeger, Kurt, Michael Martini, Jonathan Rasouli, and Anthony Costa. "Emerging blockchain technology solutions for modern healthcare infrastructure." *Journal of Scientific Innovation in Medicine* 2, no. 1 (2019). 1–7.
29. Zheng, Zibin, Shaoan Xie, Hong-Ning Dai, Xiangping Chen, and Huaimin Wang. 2018. "Blockchain challenges and opportunities: A survey." *International journal of web and grid services* 14, no. 4 (2018): 352-375.

CHAPTER 10

IOT AND AI INTEGRATED BLOCKCHAIN SECURITY

TEJASWINI R. MURGOD[1],
S. MEENAKSHI SUNDARAM[2]
[1]ASSOCIATE PROFESSOR, DEPARTMENT OF AI AND ML, BNM INSTITUTE OF TECHNOLOGY, BANASHANKRI, BENGALURU, INDIA
[2]PROFESSOR AND HEAD, DEPARTMENT OF CSE, GSSSIETW, MYSURU, INDIA

Abstract

In recent times, the Internet of Things (IoT), artificial intelligence (AI), and blockchain technology have emerged as some of the leading innovations. A blockchain is also known as a distributed ledger, which stores the complete information of all the assets. Blockchain provides a shared, distributed, and decentralised ledger that helps increase the security, transparency, and trust of business processes. Blockchain sets up a standard for the data by setting up a harmonised digital platform. It allows all the data to be stored in one format by using the hash function, which also leads to an increase in data management. IoT helps in the automation of business processes in industries where AI optimises the outcome of the business process by providing insight into the business process. These two technologies drive data management and the automation of business processes.

Blockchain technology identifies IoT network-connected devices and authenticates these devices, which will lead to an increase in trust. It also helps identify the different devices that are connected to the IoT network. Blockchain assigns a digital identity to the device based on its physical location. IoT device applications that are equipped with blockchain technology permit users to restrict the data collected by the IoT devices. Blockchain technology is transparent and easily traceable, which helps monitor the shipment of goods from source to destination in a supply chain. Blockchain technology involves smart devices that actively validate the processes and the data transactions.

Blockchain technology allows IoT-connected devices to communicate directly instead of through a centralised network. This also reduces cyberthreats. Integrating IoT with AI helps to solve various issues that deal with the large data volumes generated by billions of IoT devices. AI can also solve some of the limitations of blockchain technology. When AI is used in a point-of-sale (PoS) machine, it allows the nodes to authenticate the transactions efficiently and at a faster rate.

Keywords: blockchain technology, Internet of Things (IoT), artificial intelligence (AI), digital identity, cyber-threats.

Introduction

Earlier Blockchain technology was only discussed in the context of bitcoins and ether, but in recent years, various non-financial use cases like digital identities, supply chain management, etc. have emerged [1]. In recent years, blockchain, AI, and IoT have emerged as the new innovations. They have the ability to create new business models, improve existing business processes, and disturb whole industry processes. Blockchain provides a distributed and shared ledger that helps increase transparency, trust, privacy, and security in business processes. It maintains the information on all the assets like a register. The information stored in the distributed ledger is primarily related to finance and identities. IoT helps the business processes be automated, and AI helps optimise the outcome of the business processes.

We often neglect the interconnection between these three innovations and use blockchain, IoT, and AI separately. However, these three technologies should be applied jointly to improve the business process. One way to interconnect these three technologies could be through the IoT, which can be used to collect data from various devices. Blockchain can be used to provide the infrastructure and rules for engaging the data, and AI can be utilised to optimise the rules and business processes. Merging these three technologies leads to significant improvements in data management and business processes.

IoT devices collect a large quantity of data from smart homes, smart machines, smart devices, sensor LEDs in various locations, smart cities, and other sources.The data collected from these devices is stored on a central server without any specific standardised data formats. Business processes or companies use different legacy platforms, which makes it tedious to extract and translate data across the platforms. Blockchain sets up a synchronised digital platform that provides data standardisation. This helps various IoT devices access data from multiple parties. Blockchain uses a hash function, which stores data in one format.

Blockchain technology provides data privacy by implementing various cryptographic algorithms. On blockchain, transactions are mainly conducted using pseudonyms or, in some blockchain systems such as Monero or Zcash, are conducted completely anonymously. The blockchain system architecture also allows data to be fully encrypted before transmission. It allows only the devices to read and write their own data via public or private keys.

Various IoT devices and machines store huge amounts of data. It is necessary to guarantee that privacy and security are provided for this data. Recently, IoT data has been transferred directly from machines or devices to the database or cloud where it is collected. The data collected from the IoT devices is not encrypted, which does not ensure privacy. To provide privacy for the collected data, blockchain technology can be efficiently used.

Blockchain technology can manage the identity of IoT devices and can also be used to authenticate IoT network devices. Identity management in blockchain refers to individual companies, but with respect to IoT, it refers to identifying IoT devices and machines. Blockchain-based identities ensure that each transaction receives a digital identity on the blockchain in line with their physical identity. Based on this digital identity, transactions between individuals and companies, between individuals and machines, or between machines can be processed efficiently at a faster rate with lower transaction costs[2].

A) IoT with Blockchain

There are various advantages that are achieved by integrating IoT with blockchain. For example, implementing a distributed and decentralised blockchain technology can handle problems that deal with single-point failure in a centralised IoT architecture. The central server is also not needed as the IoT devices communicate with each other. Blockchain also provides a secure computing environment as it implements hash functions, cryptographic algorithms, and timestamps. Blockchain also provides a tamper-proof and immutable ledger. It does not allow the data to be modified in the ledger if the majority of users do not validate it. This delivers a trusted system where only the participating devices can either accept or discard the transaction.

The developments in IoT have led to numerous opportunities in different domains. IoT enables the linking of virtual and physical objects over the internet. This unpredicted development of the IoT has unlocked various innovative changes in numerous fields. However, there are still numerous issues that need to be handled with respect to IoT objects. One of the issues is the lack of trust and confidence among IoT devices. Currently, a centralised third-party authority is responsible for collecting and processing the data from various IoT devices. This centralised authority does not pose any restrictions on how the data is collected. Here, the central authority is like a black box that does not provide complete insight into data collection and processing to the IoT users.

Blockchain technology provides an autonomous, distributed, and decentralised environment. In the centralised model, there are several issues raised with respect to single points of failure, security, and trust. Blockchain technology uses a decentralised approach, which eliminates a single point of failure. It also provides security, trust, and data integrity through its tamper-proof and immutability features. [3]

Applications of integrating IoT and blockchain

Figure 1, given below, depicts the applications of IoT and blockchain integration. The areas that benefit from IoT and blockchain integration are smart homes and smart cities. Smart homes include confidential and personal information that is collected from various sensors and cameras located in the homes. Integrating blockchain with IoT can further provide security to the collected data so that any unauthorised users cannot modify the personal data.

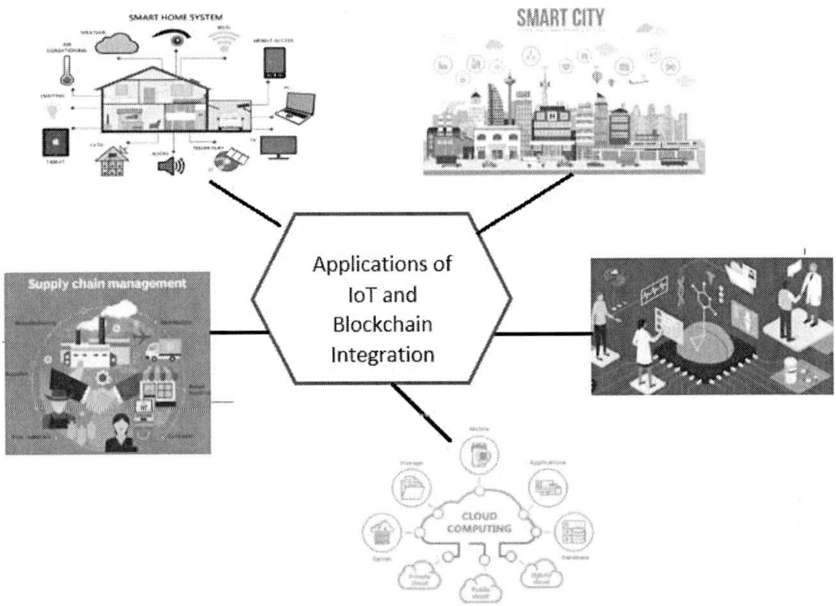

Fig.1Applicationsof IoT and blockchain integration

Supply chains also benefit from IoT and blockchain integration. Supply chain networks consist of multiple stakeholders and raw material providers. As the supply chain is complicated, it is difficult to manage and monitor the complete supply chain's activities from start to finish. As many stakeholders are involved in the supply chain, multiple delays are included, which is challenging to handle. Nowadays, companies are equipping their vehicles with IoT to track their movements during transportation. Integrating IoT with the blockchain can increase the traceability and reliability of the supply chain. The IoT motion and temperature sensors collect accurate data from the vehicles and provide details for the transportation process. Once the data is collected from these sensors, it is stored on a blockchain. The stakeholders in the smart contracts can access this real-time data and take the necessary measures.

Cloud memory is one more area that is benefited by the IoT and blockchain integration. IoT devices use the cloud to store all the information collected from the devices. There is an increase in processing power, and stored data is secured in the cloud. The smart contracts automatically perform the desired operation under the agreed terms and pay the provider once the operation is completed and delivered.

Medicine and pharmacology are the most benefited by the IoT and blockchain integration. "Counterfeit drugs" are one of the increasing problems faced in the pharmaceutical sector. The pharmaceutical industry is whole and solely responsible for the production and distribution of the drugs produced. It is difficult for the pharmaceutical industry to monitor the entire distribution of drugs. Blockchain technology is transparent and traceable, which helps the pharmaceutical industry monitor the shipment of drugs from origin to destination in a supply chain. Recently, the medical industry has been using smart medical devices. These devices assist in monitoring the complete patient's treatment process, right from collecting the patient's information and forwarding it to the doctor to providing nurse assistance when required. As these medical devices handle the patient's private information, any intruder changing the data may pose serious threats. Integrating IoT with the blockchain restricts end users from accessing all the data collected by the devices. Using blockchain technology in medicine prevents hackers from hacking medical data and also provides accurate patient information[4].

How IoT can leverage blockchain technology

The key challenge faced by the IoT is that blockchain is emerging at a faster pace. One key area of IoT development is security. Here are some of the key security challenges for IoT infrastructure and services:

As IoT devices are growing exponentially, it is really difficult to identify, secure, and authenticate them.

- Setting up a centralised security model is difficult and very expensive to scale.
- An intruder may introduce a single point of failure, and a DDoS attack can be initiated.
- Setting up a centralised infrastructure will be difficult for the industries where nodes are spread.

Security problems of the Internet of Things (IoT)

In this section, we discuss some of the common security problems of the IoT and their solutions.

1. *Difficulty in the Ecosystem*

The IoT is a broad and diverse ecosystem that integrates communication, devices, people, and interfaces. IoT untangles industrial production, but the application is not simple as it includes multiple components in the ecosystem. The different things included in the ecosystem are the various sensors laid out, the different networking devices like bridges, routers, gateways, etc., the different network and communication protocols, and the regulations that govern security and confidentiality.

2. *Restricted device capacities*

Multiple IoT devices have limited memory, power, and computational capabilities, which results in a lack of ability to implement advanced security patterns. As the advanced security features are not implemented, there are chances for an attacker to attack these devices. So the IoT architecture of the equipment must be scalable so that it can incorporate the necessary security features.

3. *Limited Experience*

The IoT technology can be relatively new, and there are not sufficient background details of the previous threats that have occurred so that we can protect the system from further failures. There is also a lack of cyber security experts specialising in IoT.

4. *Threats and attacks*

Special computer programmes are designed specifically to attack the IoT ecosystem; they are called "malware. These malware programmes take unnecessary actions without the user's consent and cause damage and theft of data. Exploit sequences are other threats that grab the weak points to access the system and will attack the infrastructure severely. Some other threats that can occur are message reproduction, message modification, network, system, or device failures, etc. The manufacturers often reduce the development time so as to release the product at a faster rate to the market. This causes a lack of fundamentals in the design phase, such as encryption and decryption techniques, access control mechanisms, etc.

5. *Privacy*

When we click the accept button while installing any application without reading the instructions or clauses given, we run the risk of losing our information. Many users just click the accept button without a proper understanding of the clauses specified. Manufacturers do not pay sufficient attention to the security breach that can occur with the system as they are busy developing and releasing the product at a faster rate. This causes leakage of user information, and the privacy of the data is not preserved.

6. *Reduced Cost*

To reduce the cost of manufacturing, companies usually limit the investment made in the security of the system. Due to this, the system may result in a state where protection is never provided. This puts end users at risk. The manufacturers usually do not consider the security issue during development, which in turn puts the end users on the hook for it.

7. *Lack of clarity in responsibility*

With respect to the security of the IoT devices, the manufacturer, service provider, and users are involved in it. The responsibilities of the key players involved with the IoT devices are not clearly stated, which may lead to conflicts when a cyberattack occurs. As the component is shared between parties, it is not clear who is managing the component among the parties. So there is a lack of clarity in the handling of responsibility.

8. *Lack of exactness in data processing*

One of the problems with the IoT is that the user is not aware of how the data is transmitted via the sensors, and he is also not aware of how the data is handled when it is transferred by these devices. The data handling mechanisms do not clearly specify how the personal information of the users is handled. During the processing stage, data may reach a third party, and the user will be unaware of it.

9. Limitation of anonymity

Users assume that anonymity is preserved with the IoT, but in reality, this is not true. In IoT, to guarantee anonymity, it is required to optimise the techniques of encryption, access control, and privacy. It is also required to safeguard the locations where the IoT devices are placed from unauthorised users.

B) IoT Security Using Blockchain

One of the problems with IoT devices is securing them from unauthorised users. The IoT market leaves plenty of room for unauthorised users to hack devices. With smart homes, smart cars, and smart cities, this lack of security can be a problem of concern. In a smart and automated car, a hacker may collect the data transferred between the IoT devices and take over a self-driving car with someone in it. In this case, security is a must and an important factor that must be provided to the IoT devices.

Biometric and two-factor authentications are some of the available security recommendations for IoT devices, but the most feasible solution could be using blockchain for IoT security. Blockchain provides strong protection against data interference. It also provides the locking mechanism for the IoT devices and allows them to be shut down. Hyundai Company has recently started up a blockchain that is specifically designed for IoT devices and network security called HDAC (Hyundai Digital Access Currency). This blockchain creates a private network for IoT device communication.

Blockchain is one of the most important concepts of recent technology. Each system consists of a small set of transactions that are executed. Each block within the blockchain maintains the details of the previous transactions in the form of a SHA-256 hash function. Creating a block is a complex step that requires multiple specialised processors and a sufficient amount of time to generate a block. As generating one block is a complex step, to tamper with the complete blockchain, the intruder needs to tamper with each and every block within the chain, which is a tedious task [5].

Cases where blockchain technology can be used with IoT

Blockchain can be used to create a secure mesh that will allow the different IoT devices to securely connect with each other and will also avoid device spoofing. Each and every IoT device is registered with the blockchain, and this will assign a unique ID to these devices so that they are recognised universally through this ID. When one IoT device requires connecting with another, it will use the ID generated by Blockchain as its URL. The device will use its local blockchain wallet while raising an identity request. The wallet will create a request that is digitally signed and then transfer it to the target device. The target device will use the public key of the sender to validate the sender's signature.

In this way, authentication will take place on the blockchain without the intervention of any centralised authority. A restricted device is allowed to connect to proxies where the wallet can be stored. This will allow the aggregation of the devices within a limited scope. This solution can be applied to various IoT services, like intelligent healthcare vehicles, transportation, logistics, etc [6].

How to secure IoT devices with blockchain

The IoT industry works on a distributed client-server model, which requires an administrator to supervise the IoT network. With respect to IoT security, the centralised authority is the weak point. The IoT sensors and devices will rely on the central authority to decide their behaviour. The security breach at the central authority will reveal the data that is transferred between the devices to the hacker. This helps the cyber hackers collect a large amount of data in one go.

In 2017, a series of these cyberattacks occurred and revealed the sensitive data of hundreds of American citizens to the hackers. The decentralised network of blockchain technology will not allow a central attack to occur on the IoT ecosystem. Hackers have to target individual nodes on the network in order to access the data that is required. The smart devices connected to the blockchain network can validate the data that is processed during the transaction, ensuring that the information is not revealed to unauthorised users.

The blockchain network would guard against the security vulnerabilities that can occur. It protects the network against a denial of service (DoS) attack and will also secure the network against any other type of irregularity occurring within the network. When a device is identified as malicious (not working correctly as per the predefined conditions of the smart contracts), it is disconnected from the other devices so that it cannot access the data transferred from other devices.

Create a trustworthy IoT environment.

To access the blockchain ledgers, users need to enter their unique code. This leads to built-in accountability with respect to all the transactions that occur within the system. The changes that are incorporated must be authorised by the users, which will help identify the users who have made the changes to the system. Unauthorised changes within the system will be restricted as the nodes attached to the network will not accept the changes suggested. For example, in the supply chain, smart devices can be used to track the movement of goods within the chain. In each step of the chain, the items are signed for authorization, which leads to tracking the movement of items in real-time as well. Any delayed or missed deliveries could be identified immediately. Implementing blockchain in the supply chain will result in increased efficiency and reliability.

Taiwanese company OwlTing realised the advantage of integrating blockchain into its supply chain and can now manage its supply chain efficiently and easily. A blockchain-based IoT system can also be helpful in the public domain, like schools and universities, where the system could monitor the appropriate time of the students' login and could also be helpful to monitor their homework status. Parents can also be involved in the school blockchain so that they can easily track their children's progress through a smart application. The application can also allow the teachers to give necessary feedback about the students. Based on the feedback provided by the teachers, parents would motivate or reward their children [7].

Prevent physical theft

Physical theft of smart devices can be tracked and secured by implementing blockchain in the IoT ecosystem. The stolen smart devices can be reported to the IoT blockchain network. The theft of the smart devices can be easily identified by the blockchain, and it then forwards this theft message to the owner of the device and to the police. The smart device location information and the secret pictures taken by the operator through the surveillance camera are also sent automatically to the police without the intervention of any third-party agency. The stolen device's manufacturing number is communicated to all the OS providers and other applications, which would then blacklist the stolen device.

B) EXAMPLES OF BLOCKCHAIN FOR IoT SECURITY

Applications that integrate blockchain with IoT will allow the devices to communicate directly through message sharing. The centralised network is not required when passing information between the devices. This decreases IoT cyber threats. According to the Gartner survey, the US has adopted IoT-based blockchain technology in the transport, energy, and pharmaceutical sectors. All these industries require the physical transportation of goods. The majority of IoT-based companies have incorporated blockchain technology into them.

One of the methods used to integrate blockchain into the IoT is by installing the chips in every IoT device. For example, the alliance of Ubirch, a blockchain-anchoring security specialist, G+D Mobile Security, and IoT carrier 1NCE have embedded chips into their IoT devices to increase the security of the IoT network. In these systems, data is not transferred from sensors to the cloud for approval. It also eliminates the 'man in the middle' attack. The device information is protected by a private key and tied up in a public blockchain. Here, the information about every access to a particular sensor is permanently recorded on a ledger. Adopting a smart contract can also be helpful to enhance cyber security.

Modum.io is a Zurich-based startup company that integrates blockchain and IoT to track the temperature of heat-sensitive packages. The logistic company's major concern is the temperature-sensitive cargo. Modum.io has developed the MODsense T temperature logger, which automatically reports the temperature every time the packet is scanned along the journey. The MODsense T temperature logger protects the customers and also allows the Swiss Post to gain insight into how temperature-related problems can be solved. The information on temperature fluctuations is maintained in the blockchain, which guarantees that this data is not manipulated.

IBM and Samsung have developed a proof of concept for a blockchain-enabled IoT system called ADEPT (Autonomous Decentralised Peer-to-Peer Telemetry). This system uses smart contracts and peer-to-peer messaging to create a distributed IoT network. ADEPT is used in the smart home. Samsung has used ADEPT to convert the washing machine into a semi-autonomous device where the device can perform self-service and maintenance. When the machine goes out of order, it will automatically notify the operator about the fault and is also capable of installing all the software updates on its own without any human intervention. The washing machine can also establish communication with other smart devices within the network to optimise energy efficiency. Using ADEPT, it can communicate and identify that the TV is on and can postpone the washing cycle until the TV is turned off. ADEPT in the washing machine can also be helpful to manage the supply of detergent it uses for washing. The machine can even place the order for the retailer to supply the required detergent. Once the retailer confirms the order, the owner gets a notification on his smart phone about the purchase order set by the washing machine. The system adopting ADEPT will soon gain much popularity in the market.

C) AI WITH BLOCKCHAIN AND IoT

AI is one of the emerging techniques in today's era. According to the Merriam-Webster dictionary, AI is defined as "the capability of a machine to imitate intelligent human behaviour"[11]. AI mainly focuses on producing new models and designs that can perform smart activities. The

AI researchers conduct multiple experiments in the lab and perform testing and validation of their hypotheses.

By integrating AI with IoT, we can produce robust technology that can solve several IoT issues. IoT devices deal with the large volume of data collected from different IoT devices. As the volume of the collected data is huge, the analysis of the data using the traditional approach is a tedious task. By incorporating AI and ML methods to analyse the data, we could extract meaningful information from the data collected by these IoT devices.

AI has the ability to analyse and provide insight from the data. This feature of AI enables multiple government and private organisations to incorporate AI for analysing the large volume of data collected by IoT devices. Existing edge/fog computing methodologies can minimise the response time of IoT applications, but by incorporating AI approaches, the system will provide better performance in a short period of time. Integrating AI with IoT can also improve the security features of the system. It also helps in identifying or predicting the external threats that can occur in a system. Users view the AI as a black box where details about how the analysis of data is carried out are hidden from them. Blockchain technology allows sharing the ledger with all the connected nodes and can help track the data's movement. Incorporating blockchain can optimise the trustworthiness of data.

Blockchain technology can also improve the effectiveness of AI by providing a secure data sharing platform and by improving the production of better AI models. When AI is implemented with smart contracts, it reduces the risk factor in the system. Smart contracts are programmed in such a way that they can perform the action only when the desired condition is satisfied. AI can also resolve some of the limitations of blockchain technology. When AI is implemented in PoW or Pos, it allows the nodes to perform transaction validation at a faster rate and more efficiently.

The mining process in blockchain requires a significant amount of energy. Incorporating AI with blockchain can reduce energy consumption, as AI approaches have demonstrated effectiveness in improving energy consumption. By adopting federated learning, AI can also resolve the scalability issue of blockchain. This is well known for providing security, but incorporating AI with blockchain can add an additional layer of security to the system.

Benefits of Combining AI and Blockchain

Some of the benefits of combining AI and blockchain include:

Authenticity:

Blockchain provides a digital record and complete insight into the AI framework. It provides insight into the data that is being used by the AI and also addresses the challenges faced by explainable AI. This helps to achieve data integrity in AI. Blockchain can be used to store the different distributed AI models by enhancing their security level.

Augmentation:

AI is capable of reading, understanding, and correlating the data at a faster rate. It can also bring the next level of intelligence to the blockchain network. This can also allow the AI to access the large volume of data that lies outside the organisation. It also helps AI gain greater insight into data usage and construct a transparent and trustworthy data economy.

Automation

By incorporating automation with AI and blockchain, we can take the business processes to the next level, where all the frictions are eliminated and speed and efficiency are increased. Example: AI models that are embedded into smart contracts can recommend expired products. It can also perform certain transactions, like reordering materials based on a defined threshold or performing payments on a specified date and time. The system can also be helpful in resolving the disputes that are raised.

Applications of Blockchain in AI

Some of the applications of combining AI and blockchain include:

- **Data Sharing:** One of the important resources of AI is data. The accuracy of AI technology is directly affected by the quality and quantity of data. Some issues may be generated while sharing the That data that is used in training the AI models is taken from different stakeholders, and it is difficult to trust these stakeholders. Authoring and validating the data generated by different stakeholders is a tedious task. Some malicious users may also share their malicious data while training the AI models. Blockchain can easily provide a solution to this problem, as there are some inbuilt solutions already in the blockchain.

- **Privacy Preservation:** One of the key issues faced in preserving privacy The personal and sensitive data of the users must be protected from unauthorised users. The data shared by the data owners must be completely under their control. Nowadays, the service providers collect the user's personal data before providing any service. This leads to the disclosure of the user's personal data to the service providers.
- **Trusted AI decisions:** AI and ML methodologies are used by different organizations. In these organisations, the entities that train the model are different from the entities that provide the data. If the data is not scrutinised properly during the training phase, this leads to improper outcomes. Some restrictions can also be included in the trained model when we employ biased data. Due to these reasons, it is difficult for the user to trust the AI model. Therefore, there is a need for a mechanism that can record the complete model creation, the data used in training, and the training. Blockchain provides a platform where users can share their data through a trusted system and also provides transparency, which is required while recording the complete process.
- **Decentralised Intelligence:** Due to the rapid evolution of the IoT, a large volume of data is generated by the devices. AI models can help in analysing and understanding this large volume of data. To build a complex training model, it is required to collaborate on the multiple devices that are distributed across various locations. There are two different methods for performing the collaboration. Firstly, the different IoT devices should share their complete data for analysis and prediction, and secondly, different IoT devices can have their own learning models and then aggregate these models for federated learning.

Companies using AI and blockchain

Some of the companies using AI and blockchain include:

CertiK:

To secure blockchain, Web3 applications, and smart contracts, CertiKprovides tools powered by AI. Certik assists in gaining insight from the data, identifying the risks involved in the system, and visualising the movement of the crypto funds. Certik also offers SkyTrace, which is a digital wallet that allows users to securely store their data in crypto currency.

Core Scientific:

Core Scientific has two substations and a dam that collects the hydroelectric power. It uses AI and blockchain to handle the processing power. Core Scientific has its own personalised AI-integrated blockchain infrastructure. It ensures that the data centres always collect optimal data and provide optimal efficiency every time they are run.

Token Metrics:

Token Metrics is a tool that analyses crypto currencies for personal investment. The techniques scan the 6,000 crypto currencies and gain market insight. It also provides suggestions for users with respect to investment decisions. Amazon, Coinbase, and Facebook are some of the companies that are using token metrics.

Figure:

Figure uses blockchain and AI integration to streamline the home loan process. It also provides new access points for consumer products, which include home improvement loans and loans taken while buying or leasing homes during retirement. Former SoFi founder Mike Cagney has an established figure and has a valuation of $3.2 billion as of 2021.

Cyware Labs:

Cyware Labs incorporates AI and blockchain-based tools into its cyber security and threat intelligence solutions. The Cyware Situational Awareness Platform (CSAP)is used to identify mobile threats, incorporate secured messages, and also report suspicious incidents. In 2021, the Global Resilience Federation and Cyware Labs joined together to create a role-based secure communication and information sharing tool.

How AI and blockchain will be the future of cyber security?

AI and blockchain are working jointly towards cyber security. The government and private organisations rely on digital systems to carry out their daily activities. This has led to an increase in hacking of these systems. As and when new technologies are used in business processes, they face greater hazards. Using newer techniques like AI, IoT, blockchain, etc. in the business process will lead to an increase in complexity. These new techniques can also become their strengths when used properly. If all these new techniques are properly incorporated, they create an additional layer of security and can prevent hackers from accessing sensitive data

2018 was a dreadful year for cyber security. Governments, national institutions, businesses, and customers faced several cyber breaches and attacks. By analysing all these breaches, it was observed that these threats occurred due to a lack of improved and properly designed protection standards. The attack was not just because of digitisation. The digitisation provided faster and better communication between the entities located at remote locations.

AI and intelligent security solutions

Nowadays, AI is used to provide solutions for security in intelligent systems. The IT company just adds the protocols or even the raw code. It is the responsibility of the AI to further add another layer of security by learning about threats and security breaches. Combining AI and ML with traditional security applications has laid the foundation for the development of new intelligent security systems that are capable of handling new threats that occur in a system.

This system is based on AI algorithms that are capable of learning about security breaches occurring in the system to prevent them from happening in the future. As the system is hit by more attacks, it will become more reliable to defend itself in the future. Daniel Miessler, Director of Advisory Services at IOActive, explains what service resilience means in the case of a hypothetical DDoS attack: "A big trend I see is a focus on service resilience, i.e., making it so that a DDoS can melt one provider or one data centre, but your service will automatically migrate to another site that can serve the same content."

Conclusion

IoT, AI, and blockchain are the leading innovations of this decade. Some non-financial uses of blockchain, such as supply chain management and digital identities, have also gained significant recognition in these years. Cloud memory is one of the areas that are benefited to the maximum extent by integrating IoT and blockchain. IoT devices use the cloud to store all the information they have collected and can also increase their processing power. The major concern is providing security as per the agreement made at an earlier stage. By incorporating smart contracts, it will automatically perform the operations according to the agreed terms and conditions.

Blockchain technology is transparent and traceable, which allows monitoring the shipment of goods in the supply chain. The applications that are equipped with blockchain and IoT will allow end users to control the access to data that is collected from various devices. The system grants or denies access to sensitive data based on the privileges assigned. Users will be assured that their data will not be accessed by unauthorised users. When blockchain is used in the medical sector, it not only restricts hacking of medical devices but also helps in gaining accurate patient history data.

The blockchain is a decentralised network. When the hacker wants to hack the data, he has to target the individual nodes that are distributed across the network. The smart devices in the blockchain network are capable of validating the data during their transactions. IoT drives industries and business processes towards automation. AI helps improve business processes by optimising the outcome. When these three techniques are merged, there is definitely a foundation for an efficient, reliable, and secure automatic business model.

Blockchain maintains the identity of IoT devices and can be helpful in authenticating them. Identity management here refers to maintaining the identity of the IoT devices and machines. Blockchain ensures that each device receives a unique digital identity with respect to its physical identity. Blockchain provides a secured computing environment by incorporating hash functions, cryptographic algorithms, and timestamps. It also provides a tamper-proof ledger to protect the data against harmful attacks. The modified data cannot be inserted into the ledger until the majority of users who are using that data validate it. This will increase trust in the system, where only the participating devices are allowed to accept or discard a transaction.

Blockchain-integrated IoT applications enable information sharing directly among the devices rather than through a centralised network. This decreases the cyber-threats in IoT. Currently, the energy, transport, and pharmaceutical sectors in the US have adopted IoT integration with blockchain at the highest rate. The smart devices can report the blockchain network as being stolen and can then forward this data to the police and the owner of the device.

By integrating AI with IoT, we can produce robust technology that can solve the problems raised by the large volume of data gathered from IoT devices. Analysing this data using the traditional approach is not feasible. AI and ML learning models can flawlessly analyse this data and extract the exact meaning from it. Various organisations have started using AI to analyse their huge data collections. Incorporating AI with IoT has significantly reduced the response time. In addition, it can also predict the external threats that can occur in the system.

As the mining process requires sufficient energy, AI can improve the energy consumption of the blockchain network. AI can also solve the scalability issue of blockchain technology. Although blockchain provides security to the system, AI adds an additional layer of security.

References

1. Rabieinejad, Elnaz, and Shahriar Mohammadi. "Integration of Blockchain and Internet of Things (IoT): The Static and Dynamic approach, Challenges, and Future Research."
2. Atlam, Hany F., Muhammad Ajmal Azad, Ahmed G. Alzahrani, and Gary Wills. "A Review of Blockchain in Internet of Things and AI." Big Data and Cognitive Computing 4, no. 4 (2020): 28.
3. Latif, Sohaib A., Fang B. Xian Wen, Celestine Iwendi, F. Wang Li-li, Syed Muhammad Mohsin, Zhaoyang Han, and Shahab S. Band. "AI-empowered, blockchain and SDN integrated security architecture for IoT network of cyber physical systems." Computer Communications 181 (2022): 274-283.
4. Guergov, Sasho, and Neyara Radwan. "Blockchain Convergence: Analysis of Issues Affecting IoT, AI and Blockchain." International Journal of Computations, Information and Manufacturing (IJCIM) 1, no. 1 (2021).
5. Singh, Saurabh, Pradip Kumar Sharma, Byungun Yoon, Mohammad Shojafar, Gi Hwan Cho, and In-Ho Ra. "Convergence of blockchain and artificial intelligence in IoT network for the sustainable smart city." Sustainable Cities and Society 63 (2020): 102364.
6. Singh, Sushil Kumar, Shailendra Rathore, and Jong Hyuk Park. "Blockiotintelligence: A blockchain-enabled intelligent IoT architecture with artificial intelligence." Future Generation Computer Systems 110 (2020): 721-743.

7. Daniels, Jeff, Saman Sargolzaei, Arman Sargolzaei, TareqAhram, Phillip A. Laplante, and Ben Amaba. "The internet of things, artificial intelligence, blockchain, and professionalism." IT Professional 20, no. 6 (2018): 15-19.
8. Alrubei, Subhi M., Edward Ball, and Jonathan M. Rigelsford. "The use of blockchain to support distributed AI implementation in IoT systems." IEEE Internet of Things Journal 9, no. 16 (2021): 14790-14802.
9. Hammoud, Ahmad, Hani Sami, Azzam Mourad, HadiOtrok, RabebMizouni, and Jamal Bentahar. "AI, blockchain, and vehicular edge computing for smart and secure IoV: Challenges and directions." IEEE Internet of Things Magazine 3, no. 2 (2020): 68-73.
10. Unal, Devrim, Mohammad Hammoudeh, Muhammad Asif Khan, Abdelrahman Abuarqoub, Gregory Epiphaniou, and RidhaHamila. "Integration of federated machine learning and blockchain for the provision of secure big data analytics for Internet of Things." Computers & Security 109 (2021): 102393.
11. Merriam-Webster. (n.d.). Artificial intelligence. In Merriam-Webster.com dictionary. Retrieved July 10, 2023, from https://www.merriam-webster.com/dictionary/artificial%20intelligence

CHAPTER 11

BLOCKCHAIN WITH AI FOR B5G WIRELESS COMMUNICATION SECURITY

VASUKI A[1], VIJAYAKUMAR PONNUSAMY[2], NEMANJA ZDRAVKOVIC[3], EMILIJAKISIC[3]

[1]SRM INSTITUTE OF SCIENCE AND TECHNOLOGY, VADAPALANI, CHENNAI, INDIA

[2]SRM INSTITUTE OF SCIENCE AND TECHNOLOGY, KATTANKULATHUR, CHENGALPATTU, INDIA

[3]BELGRADE METROPOLITAN UNIVERSITY, SERBIA

Abstract

To enable connectivity for an exponential number of users, allow multi-gigabit transfer rates. Beyond the recently formed Fifth-Generation (B5G) or 6G networks, the forthcoming 6G technology will meet the demand for ever-increasing industrial services and applications. This technology combines diversified resources and communication protocols for flawless access. Hence, security is the primary concern in 6G networks. B5G handles hundreds of additional bytes daily from the Internet of Everything and offers a transmission capacity of up to terabits. Due to its high capabilities for content distribution, caching, and processing, the transition from 5G to B5G is expected to aid the Information-Centric Network (ICN) significantly. The authenticity of content within ICN is crucial since security could be a problem in B5G. The above-mentioned concerns about privacy and security can be resolved by utilising blockchain technology. Blockchain can provide an underlying, transparent, and secure environment for networking within B5G. Blockchain uses a distributed ledger to provide a secure service without the need for third-party support. This chapter reviews the security concerns within the B5G network and how blockchain and artificial intelligence (AI) can enable a robust privacy framework.

Keywords: Blockchain, AI, B5G networks, Communication Security

Current challenges within B5G networks and their need to be secure

The evolution and significance of B5G networks are briefly covered in this section. Recent research studies' findings on security concerns and solutions are highlighted. This section discusses how B5G will alter communication, the industries it will affect, and how to get ready for upcoming security concerns.

Introduction of B5G networks

B5G networks in both academia and business are driving a digital revolution in telecommunications. B5G networks are the newest mobile data transmission standard. Unprecedented new advanced types of data transmission will also be made available, in addition to speed gains. A more fundamental realignment must occur for businesses to benefit from B5G rather than merely look to the technology for faster downloads.

The B5G network is created to attain broad spectrum, efficiency, computation, and mass networking dueto immense growth in the Internet of Things (IoT)devices. These devices will enable the delivery of updated services like sharp transit, environmental impact analysis, telemedicine, AI, and digital triggers. This transmits full High-Definition (HD) video to active drones and robots. Existing multiple-access systems struggle to support the current number of devices in existence, and this is predicted to increase to 25 billion by 2025.

Wireless communication technologies have been vulnerable to security threats since their inception. The security issues reported in the earlier wireless technologies are listed in Table.1.1.1.and illustrated in Fig.1.1.1.

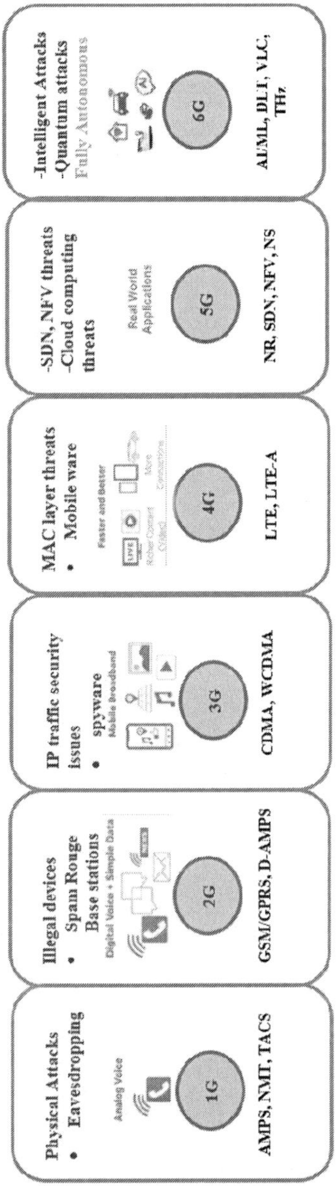

Fig.1.1.1. Evolution of network security in wireless technologies

Wireless Technology	Security threat reported
1G	Wireless channels are used for unauthorised copying and masquerading.
2G	Message spamming
3G	Security flaws that were present on desktop computers have spread to wireless devices.
4G	Due to the numerous third-party programmes, the threat landscape has become complex and dynamic.
5G	User's perspective Location tracking, identity, and other personal data are the main privacy concerns. 5G antennas - Users inside and outside can be precisely tracked using 5G antennas. International Mobile Subscriber Identity (IMSI) - An attacker can monitor someone's activity by intercepting mobile traffic in a specific area and obtaining the IMSI of the subscriber's device.
4G and 5G	The 4G network technology offers a vast coverage area because only one cellular tower distributes the signal. Compared to 4G, the coverage area of 5G networks is significantly smaller, and the signal does not penetrate barriers either. As a result, 5G networks require both indoor and outdoor base stations with numerous small antennas.

Table 1.1.1 Security threats in wireless networks

Wireless Technology	Security threat reported
1G	Wireless channels used for unauthorized copying and masquerade
2G	Message spamming
3G	Security flaws that were present on desktop computers have spread to wireless devices.
4G	Due to the numerous third-party programs, the threat landscape has become complex and dynamic.

5G	• User’s perspective Location tracking, identity, and other personal data are the main privacy concerns • 5G antennas Users inside and outside can be precisely tracked using 5G antennas • International Mobile Subscriber Identity (IMSI) An attacker can monitor someone's activity by intercepting mobile traffic in a specific area by obtaining the IMSI of the subscriber's device
4Gand5G	• The 4G network technology offers a vast coverage area because only one cellular tower distributes the signal. • Comparedto4G, the coverage area of 5G networks is significantly less, and the signal does not penetrate barriers either. • As a result, 5G networks require both inside and outdoor base stations with numerous small antennas.

Security challenges in B5G networks

Compared to the currently built 5G network, B5G applications will pose strict criteria and need expanded network capabilities [1–2]. They can be grouped by the vast range of major B5G use cases for which they were created. They have a significant impact on the implementation of B5G security as well.

1. Enhanced Ultra-Reliable, Low Latency Communication (ERLLC) - The effect of latency in security workflows is considered for better servicing ERLCC. Authenticity requires extremely secure measures to ensure that resources and services are available.
2. Enhanced mobile broadband (FeMBB)
3. Extremely high data rates can make it impossible to process security-related traffic, such as intrusion detection, AI/machine learning (ML) processes, transit examination, and enciphering.
4. Ultra-massive Machine Type Communication (umMTC) for effective communication.

Critical use cases will be served by umMTC, which has even stricter security requirements than 5G. The Internet of Everything (IoE) will most likely be used to evaluate the implementation and functions of secure outcomes like dispersed AI and ML and confidentiality issues. Fig. 1.1.2 shows the basic security challenges in B5G networks.

Solutions to security challenges in B5G networks

Numerous security measures were employed in the 2G, 3G, and 4G generations of communication networks [3]. For example, in the physical layer of 2G–4G networks, hybrid automatic repeat request (HARQ) techniques that combine forward error correction (FEC) and automatic repeat request (ARQ) channel codes are extensively used to locate faulty inputs and rectify them for data validation support. In addition, to ensure information reliability and identify data transmission, repository, and contraction problems, computation techniques such as Cyclic Redundancy Check (CRC) have been applied in the Radio Link Control (RLC) layer. However, it appears that the security methods and designs used in preceding generations (2G–4G) are not compatible with 5G and B5G networks for the following reasons:

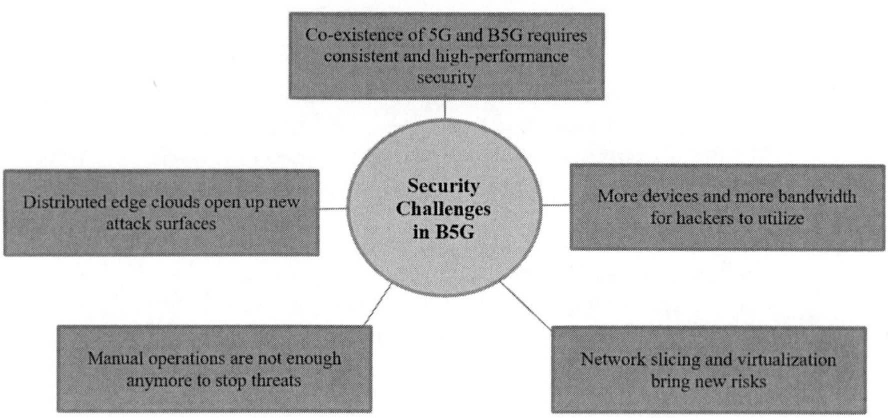

Fig.1.1.2 Security challenges in B5G networks

- An essential factor is that the security measures mentioned above, used in 2G and 4G networks, cannot solve the problem of data tampering, including deletion, injection, and interleaving in 5G and B5G.
- A further aspect is the changing nature of emerging services and technologies in B5G networks. In addition to maintaining data integrity, these technologies and services place new demands on security and integrity.

In the case of B5G networks, due to their advantages, they are creating new exemplars in real-time scenarios. But that still leaves a few issues with security and privacy recreating new exemplars in real-time scenarios. But that still leaves a few issues with security and privacy. Public databases called blockchains provide an alternative to centralised systems that have historically been used in numerous fields, such as manufacturing, economics, and industry. Decentralisation, transparency, immutability, and other desired qualities of blockchain make it a potential technology for resolving security challenges. In [4], it explores common safety and confidentiality issues in edge intelligence in B5G networks and develops a structure for connecting blockchains of such systems, which can guarantee both privacy and confidentiality.

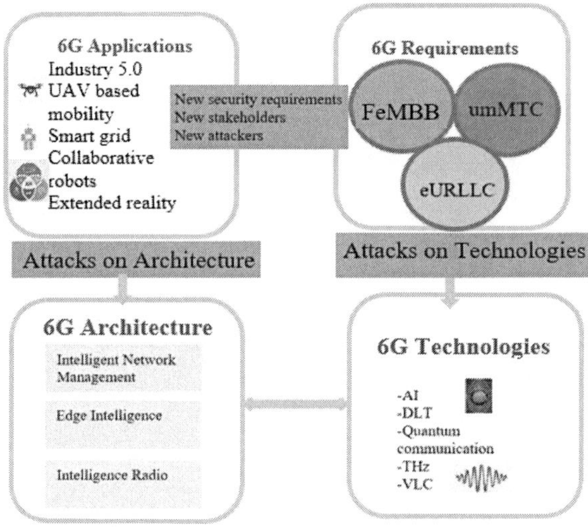

Fig.1.1.3 Security composition of 6G networks

Impact of blockchain on privacy enhancement within the B5G network

To enable smooth access to the services, B5G technology combines a range of heterogeneous resources and communication protocols. Due to its diversity and authorization delegation, the 6G communication ecosystem prioritises security. Asymmetric cryptography is used by blockchain to provide security with high levels of authenticity, integrity, and non-repudiation.

Blockchain-based smart contracts can offer data privacy features such as data provenance, access control, and auditability. This section discusses how blockchain can help maintain and grow a secure B5G network. The role of blockchain is illustrated in Fig. 1.2.1.

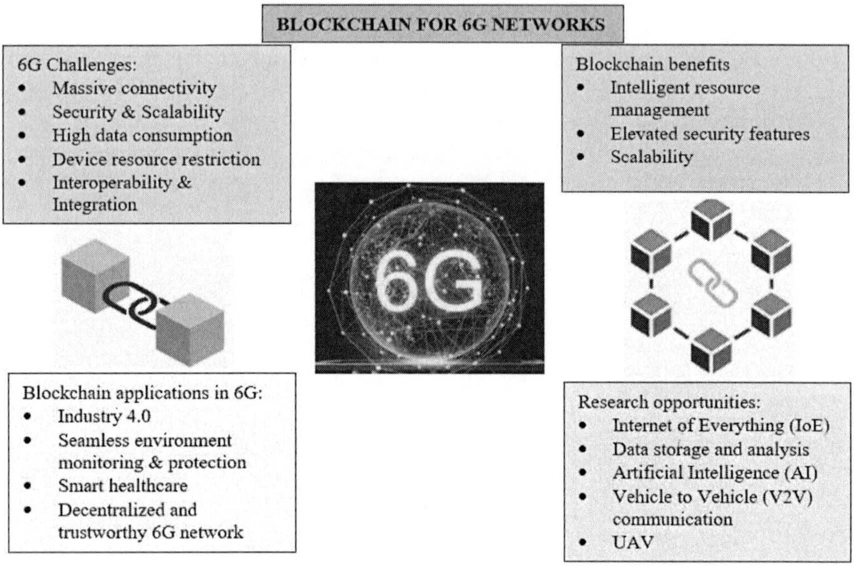

Fig.1.2.1 Role of blockchain in 6G networks

Blockchain

Blockchain can identify several safety and reliability problems in communication networks, including improving reliable data communication, enabling more efficient information sharing, and providing additional features-secure access control, validation, and management capabilities for today's 5G and B5G networks.

A chain of information structures makes up the blockchain. Every piece of information creates a square, which is connected to a series of requests. Blockchain is a cryptographically secured distributed ledger technology that cannot be altered or tampered with. The general architecture of blockchain is illustrated in

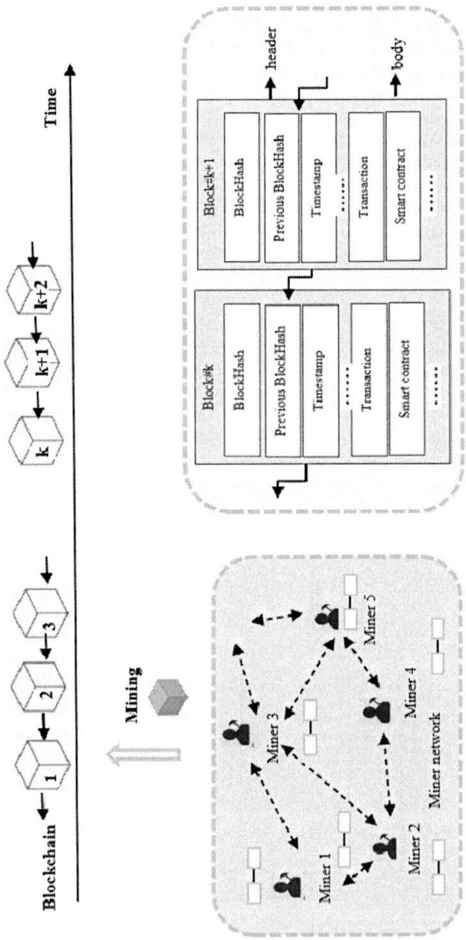

Fig.1.2.2. Basic Structure of Blockchain

Blockchain terminologies

There are key components in a well-characterised blockchain:

- The depicted registry: Tamper-proof records are kept in the blockchain's central ledger.

- Cryptography: The keenness of the record, the veracity of the record, the mystery of exchanges, and the personalities of the members are all secured utilising public keys.
- Consensus: A distributed mechanism is used to decentralise system administration. A consensus protocol specifies how a transaction is approved and added to the blockchain systems. Then, consensus and block generation time are connected. The consensus method is categorised into two types: evidence-based consensus and voting-based consensus.

Miner nodes must use evidence-based consensus mechanisms to indicate which peers are more authentic than others and beneficial to adding advanced transactions to the blockchain.

- Voting-based consensus techniques are widely used in collaborative Some node decisions affect how blocks are structured. A voting-based system can use a fully integrated topology between these nodes to confirm blocks and reach a final consensus. This contrasts with Proof-of-Transfer (PoX), which relies on poor internet access to assure blocks and attain resolutions.
- Proof of Work (PoW)**:** The rivalry of node processing power is the central concept of the PoW algorithm, which was first put forth in the context of Bitcoin. Miners can clean up their books by expending a lot of computational power to find hash values that meet their need. After the other nodes have verified the block, it will be included in the blockchain. The bookkeeping node will then receive a reward. A hostile node takes a lot of effort to turn off the PoW consensus mechanism (control over 50% of the nodes). As a result, it can restrict the misbehaving nodes' bad behaviour. Decentralised PoW allows nodes to freely enter and exit. However, it will undoubtedly result in resource waste and poor efficiency.
- Proof of Stake: The PoS algorithm solves the resource-wasting problem of PoW. The quantity and duration of tokens that each node must mine make mining less challenging. It reduces the time needed to obtain consensus and helps the PoW algorithm avoid wasting many resources. PoS advantages affluent miners, but it also has the potential to create a monopoly. Blockchain projects adopting PoS algorithms often run a PoW consensus process for a period of time before moving to PoS to avoid concentrating stakes on a small number of nodes.

- Incentive: Rewards for various contributions to the system are determined by incentives. Users of blockchain systems are rational. As a result, motivation is one of the most important business logics in the blockchain.
- Smart contracts: Blockchain uses smart contracts and cryptocurrency programmes that make it possible to build multi-step processes to carry out automated transactions. Network members automatically implement and carry out the existing contracts without the help of a third party or a central node once a contract is activated and registered in the blockchain. By integrating security considerations into the standard block-generating phase, privacy-preserving safe block generation is accomplished. The smart contract is called whenever a new task is created to verify the pertinent data. A task threshold is used to express the pertinent information.
- Additionally, the newly received task's security level is being compared to the security threshold. The task is bonded into the block if the necessary security levels are met; otherwise, the task will be checked with lower-level The competence of the task is assessed as the task receiver checks for it, and it is compared to the block's security level. The Secure Hash Algorithm (SHA-3) hashing technique determines the hash value.

Blockchain-enhanced B5G networks

As shown in Fig. 1.2.3, blockchain technology facilitates more efficient sharing of resources, enhances trusted data interactions, strengthens secure access controls, protects privacy, and facilitates tracking, authentication, and surveillance. It is capable of figuring out multiple trust and security-related problems in communication networks by providing functionality for current 5G and upcoming 6G networks.

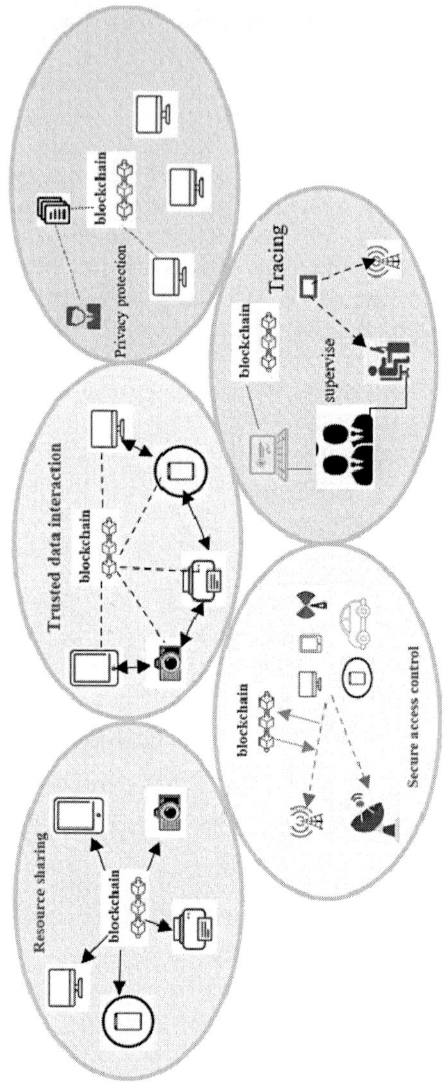

Fig.1.2.3. Applications of blockchain

High-level security should be provided for 5G networks using decentralised ledgers. By offering distributed trust models with high access authentication, blockchain contributes to the security of B5G networks. It allows B5G systems to defend and assure information secrecy. It is much harder for intruders to deal with input when it is reserved as data information across a network of computers (i.e., IoT metadata) [5]. Additionally, smart contracts can support B5G services such as data authentication, user verification, and protection of B5G resources from threats in their capacity as a trusted third party.

i. Blockchain in Resource Pooling

Isolation of resource hosts, which may be lacking motivation or have cost and security concerns, actually discourages the pooling of resources and makes collaboration and coordination of network entities impossible. The categories and amounts of network resources, such as computing and storage resources and network slices for new 5G systems, cloud processing, multi-access edge computing (MEC), software-defined networking (SDN), and network function virtualization (NFV), complicate the management and sharing of resources. By efficiently promoting collaboration and alleviating concerns about security and trust among scattered network entities, blockchain and its intrinsic properties can effectively foster resource sharing.

ii. Blockchain in spectrum allocation

Intensive research has been carried out on blockchain applications for spectrum sharing. The recent research works that are focused on blockchain in spectrum allocation are listed in Table 1. 1.2.1.

Table1.2.1Blockchaininspectrumallocationanditssecurity

Research work	Role of blockchain
[6]	Presented a method for operators to share a spectrum that is based on a consortium blockchain and that provides reliable privacy and security guarantees.
[7]	A blockchain authentication method was proposed to enable and ensure the sharing of resources (spectrum) in cognitive mobile networks without continuous spectrum control.
[8]	A blockchain-enabled spectrum allocation framework has been implemented, which can convince key users to share unused spectrum and ensure successful spectrum allocation with minimal effort.

[9]	A semi-distributed wireless network's unlicensed spectrum management framework is built using blockchain to address spectrum conflict.
[10]	Presented an intelligent network design using smart contracts to handle the unlicensed spectrum sharing between operators and users

iii. Centralized network to decentralized network

Blockchain offers a safe and effective method for managing diverse infrastructures and devices in 5G and IoT networks. Blockchain can eliminate the requirement for third-party authorization by decentralizing the network architecture and doing away with the mechanism of centralised network management. For example, blockchain-based cloud computing enables the decentralization of cloud/edge B5G networks by removing centralised control of the core network and providing decentralised fair agreement with the blockchain consensus platform, removing bottlenecks of single point failure, and significantly upgrading the system's confidence.

Additionally, creating an end-to-end network using blockchain might increase the security of device-to-device (D2D) communication by converting each D2D device into a node that can validate and monitor transactions for greater system transparency and dependability [11]. This section introduces Selective Blockchain (SBC), which is based on the Selective Byzantine Fault Tolerance (SBFT) consensus algorithm to select cooperative and standard verifiers based on the device's security level.For the D2D nctwork's optimised blockchain systems, the SBC consists of off-blockchain and on-blockchain structures. It explains the overall architecture and how to choose and authenticate specific nodes that want to participate in the consensus procedures.

The general architecture of the SBC employing the SBFT method outlined in [11] is shown in Fig. 1.2.4. There are two basic components to the SBC. Low-power IoT device verification and user data gathering are done off-chain before sharing and storing the data for D2D communication. Users can gain the necessary rights and join the blockchain network by using the device-to-device communication protocols, referred to as on-blockchain, in the other region of the SBC. The blockchain's consensus algorithm, Smart Registration Services (SRS), and SBFT are used to assess and manage the ecosystem's or the blockchain's security.

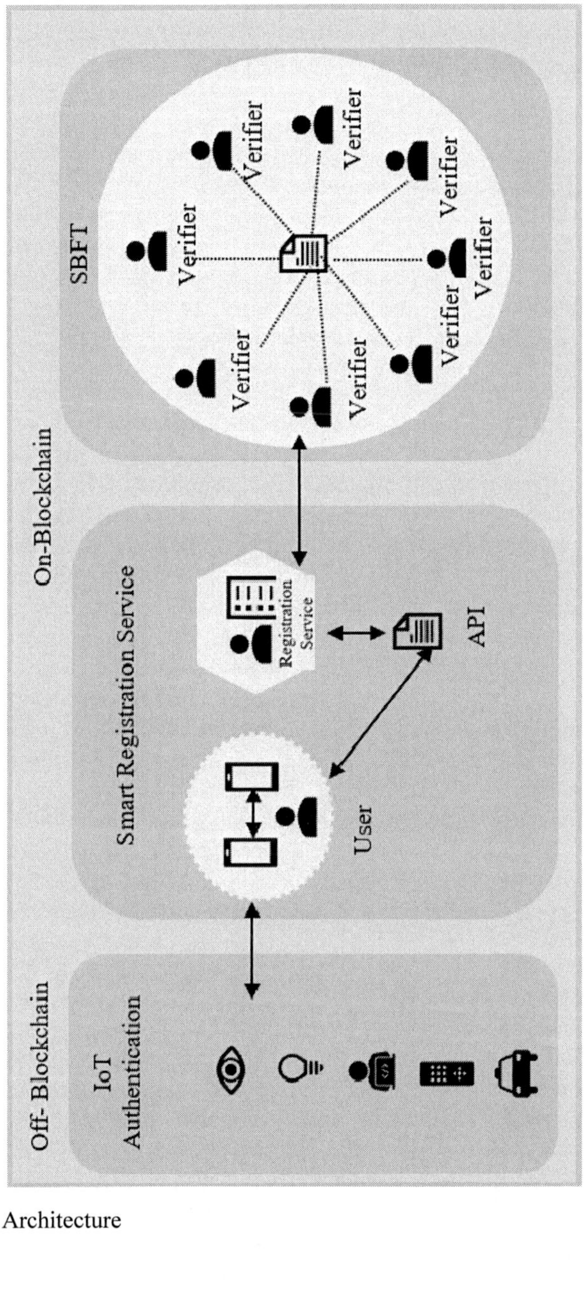

Fig.1.2.4. SBC Architecture

Role AI integrates blockchain to enable secure B5G network

B5G networks need to know how to adapt to extremely complex network topologies and adequate memory and processor implementations to be more effective and stable. Blockchain technology and AI are anticipated to be powerful techniques for the viability advantage of present and future B5G or 6G networks, according to various Quality of Service (QoS) standards. Cellular mobile network operators have new issues as 5G takes off; therefore, the company attempts to address this complexity by integrating AI technologies into network systems.

Implementing AI in network systems aims to lower capital expenditures, enhance network connections, and foster inclusive economic growth. B5G network architecture could experience a breakthrough in terms of uninterrupted connectivity, low cost, excellent privacy and security, and higher productivity by integrating blockchain and AI. As it uses a common system to deliver dependable results even without a private organisation's involvement, blockchain can provide a B5G-enabling private and transparent communication infrastructure. The processing, transmission, and retention capacities of B5G systems could be considerably improved by using AI algorithms to identify the optimum qualities for blockchain technology. Additionally, a variety of applications, such as telehealth, smart manufacturers and cities, industrial automation, remote monitoring, and unmanned aerial vehicles (UAVs), are studied using the most recent developments in AI-enabled blockchain systems in 5G networks. Combining AI and blockchain technology will undoubtedly significantly increase service quality and customer pleasure. AI and blockchain can do away with the centralised network organisation model by decentralising the network architecture and removing the requirement for a third-party authority.

In the future, AI and blockchain for B5G networks will present more significant opportunities for the digital communications sector because they might be adapted to attempt to engage end-users further and minimise new implementation and conservation. Blockchain algorithms will be used in aggressive cyberattacks against AI that involve altering a product's aspects to trick a system into classifying it differently than it is. Threats like decentralisation, secrecy, data system integration, transmission protection, and system weaknesses are particularly vulnerable to blockchain technology. This blog promotes the use of AI-assisted blockchain systems by researchers and innovators to improve data protection, mainly after they are implemented in the public sector and cause 5G and future generation networks to be more dependable, safe, and efficient.

Due to the developing tensions between security and privacy, it became a conflict or propensity to adopt blockchain technology, which is generally used to store information securely on a distributed system. This vast data can be useful in a variety of applications by applying ML techniques to it. Systems containing large or vast amounts of data can benefit from using ML models. In summary, blockchain and ML can be combined for the betterment of the future. This section focuses on the benefits of using AI and blockchain to enhance confidentiality and safety in B5G networks.

Blockchain technology limitations in the B5G network

Various security threats against blockchain are unavoidable. Some potential attacks are discussed in the following sub-section [12].

Alternative history attack

Double-spending: When a user alters a blockchain network and inserts the authenticated code, it allows the user to purchase another coin. Although it allows this process, there is also a probability that coins would be detected for those which were not enciphered properly.

By secretly mining a different fork, an attacker could use different incursion techniques for double-spending, exposing the vulnerability in distributed systems. If the attacker is fortunate enough to produce a longer false chain, it can invalidate a confirmed chain that the mining network has accepted. If an adversary has more than 50% of the network's computing power, it can successfully carry out a 51% assault, also known as a Goldfinger attack, in a PoW system.

Selfish Mining

Selfish mining's main goal is to increase the likelihood that an attacker will succeed by allowing honest miners to use more power on the public chain. An attacker keeps mining after discovering a legitimate block without releasing the recently generated block. Until other mineworkers discover a substantial piece, the aggressor will broadcast all already mined pieces to the organisation.

Cryptanalytic attack

Theoretically, cryptanalytic attacks against blockchain focus on weakening the encryption code and revealing its keys. The blockchain's underlying technology and cryptographic algorithms are interconnected. Because of the unpredictability of key creation and the vulnerability to private key leaking, key attacks typically occur undetected.

Nothing at stake

PoS and other affordable alternative consensus mechanisms are significantly more exposed. When anything is in danger of being lost, it is in danger of being lost; but when nothing is in danger, the adversary tries to launch "attacks with nothing to lose" and work on numerous branches simultaneously since they have nothing to lose. In this way, an attack with nothing at stake takes shape.

Conventional cyber-attack:

Blockchain is susceptible to typical intrusions comprising man-in-the-middle, distributed denial of service (DDoS), replay, eclipse, Sybil, and man-in-the-middle attacks. When several blockchain nodes are subjected to a DDoS attack due to an influx of incorrect requests, they may abruptly stop functioning normally. In contrast to man-in-the-middle attacks, where intruders could insert a new input, the replay attack entails intercepting the communication partners' data packets and sending them to their destinations unchanged.

Although powerful, blockchain has certain flaws that make it vulnerable to the majority, or 51%, of attacks.Therefore, it is necessary to cover up blockchain by integrating it with AI [13].

Integration of AI and Blockchain

This section includes a quick discussion of how blockchain and AI are combined.

- AI in Blockchain: AI can conceal blockchain in the following ways:
- Energy consumption: Mining is one of the blockchain's ideas. Mining is a challenging operation that requires plenty of resources is a challenging operation that requires plenty of resources. It has been shown that AI is quite effective at maximising energy use. If

we think about efficiency, we can maximise the energy for the blockchain. AI will assist in lowering the price of mining.

- Security: Although it is nearly impossible to hack the blockchain, we later discover that the layers and implementations are not very secured. Recently, AI has been progressing fairly quickly. AI will enable blockchain to ensure the secure deployment of apps, particularly if the system's reliable structure is preserved.
- Efficiency: The whole cost of authenticating and disseminating transactions on the blockchain, according to a 2016 Deloitte estimate, would be $600 million. In this case, an intelligent system might be When other miners are alerted to halt working on a transaction, they may instantly assess which nodes are most likely to complete it first, lowering the overall cost.
- Blockchain in AI: A blockchain can benefit AI in several ways, such as increased machine productivity due to fewer errors or reduced attacks, lower costs, quicker mining, etc.

Blockchain can hide AI in the manner listed below:

- Can aid AI in defining itself: AI black boxes have a problem with explanations. It is powerless to defend If there is a clear path, the data and the models will be more trustworthy. The computer's decision-making process can also be observed. As a result, if any decision does not convince us, we can analyse and evaluate the procedure independently.
- To make AI more practical, it is more efficient because there is more data for AI to evaluate. Thus, secure data exchange generates more data, which generates more models, which generate better behaviour, better results, and more fresh data.
- Reduced risk of disaster scenario: An AI that has been programmed with certain smart contracts in a data access object could only perform the defined operations, such as the execution of digital contracts.

Blockchain and AI-enabled B5G networks: security and privacy

This section outlines our concept for combining blockchain technology with B5G, which is AI-enabled. We present a conceptual framework and argue how blockchain can help AI build a smart and secure B5G network based on recent research directions.

In [14], the work explores the security of blockchain-authorised data processing for AI applications. The proposed secure architecture for an indoor navigation system is illustrated as shown in Fig. 1.3.1. The main security issue, namely, data storage security, which is present in AI applications, is addressed by constructing the blockchain as the repository platform as a consequence of its rigid and non-pirated nature.

Fog server: Due to its powerful computational capabilities, a fog server can be utilised for outlier identification, job management, and user administration. The fog server operates the consensus method in the blockchain ecosystem as a computing node in the interim. Additionally, the fog server hires reliable personnel based on influences that can significantly raise the standard of path utilities.

Blockchain: Blockchain technology links workers and requesters and has decentralised information-storing capabilities. On platform transactions, requesters and employees could concur.

Trusted Authority (TA): TA manages key secrets which can be used to allow legitimate registration identities for taking part users without exposing their privacy. In particular, the inclusion of TA enables the blockchain to actively carry out unrestricted task transactions and non-tamperable data storage.

Users: Requesters and employees are two distinct participant user categories in the navigation system. The people who desire to know their location are referred to as requesters. They send the platform arbitrary location queries using smartphones or other portable electronic devices. After each service, they assign employees equivalent feedback scores based on the quality of the service. Workers are accountable for delivering results to requesters (i.e., path information).

A smart contract can represent the requester-worker link in the crowdsourced interior navigation system powered by blockchain. A smart contract is established when a requester publishes information about their location requirements and the blockchain platform. Here, we assume that the requestor uploads their request using a private key and the employees do the operation using a public key. The smart contract runs a validation function when a worker wishes to respond to a request to see if the worker's reputation value satisfies the requirement. A qualified individual can carry out a task and adhere to the terms of the contract.

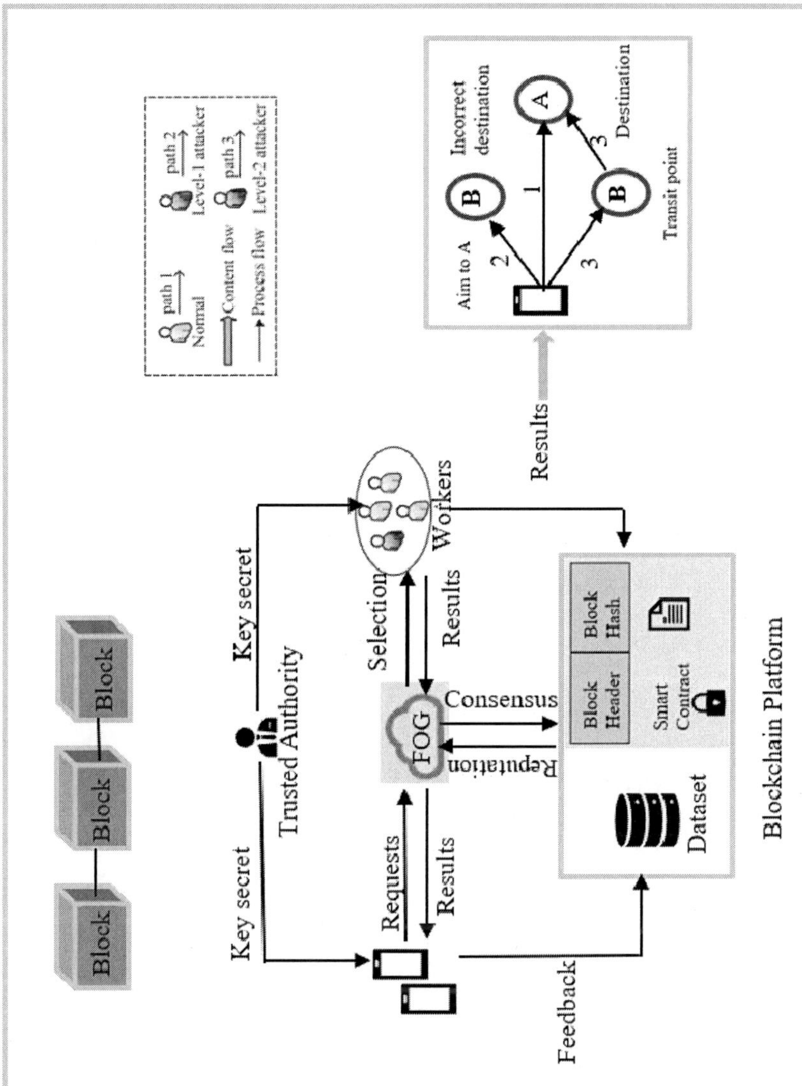

Fig.1.3.1. Blockchain-based indoor navigation system

The blockchain platform and the fog server perform the reputation management algorithm and the path-detecting algorithm. The fog server assesses the path's aspects, such as whether it is the shortest route, while using the path detection method. Based on the outcomes of path detection, a worker's reputation value is automatically updated after a task is completed.

According to Fig. 1.3.2, using blockchain and AI, the study's authors [15] look into the material to trust the ICN for B5G. It is primarily suggested to use TrustCoin, a blockchain-based system for protection validation and transmission, to precisely and dynamically assess the dependability of B5G nodes. Based on the subjective and qualitative nature of B5G nodes, deep reinforcement learning (DRL) is suggested to evaluate the trustworthiness of the material. A novel incentive mechanism eventually builds domestic certainty in B5G networks. The experiments' findings show that the suggested trust-ICN architecture and process in B5G, which AI and a blockchain bolster, are workable.

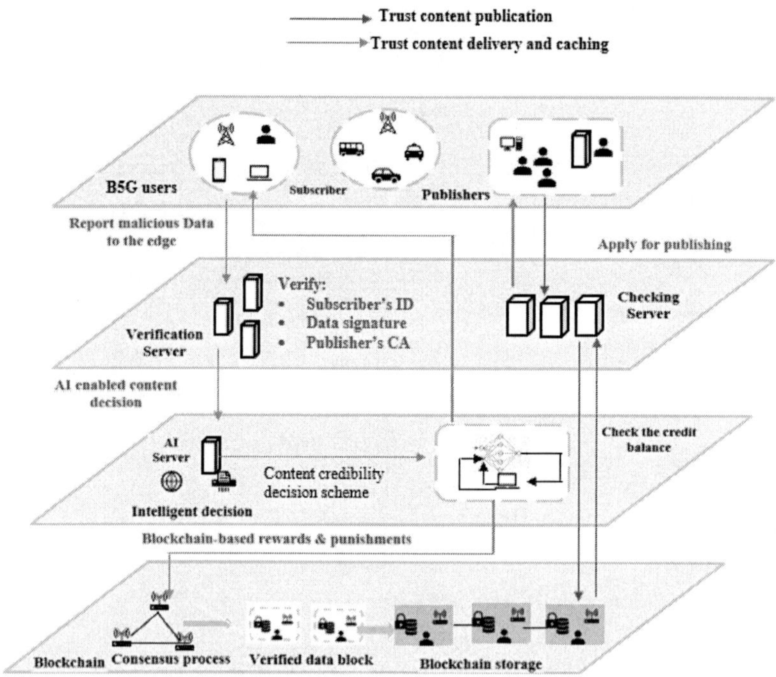

Fig.1.3.2. ICN-Trust architecture

The study's authors [16] note that the imbalance between growing connectivity and limiting CPU capacity could enable attackers to make DoS/DDoS attacks more severe. By creating a blockchain-based incentive structure that recognises DoS/DDoS attacks and penalises similar activity by imposing excessive transit in communication, this study seeks to minimise the impact of DoS/DDoS assaults. Attackers are blocked from approaching secured servers without inspection by the defensive mechanism by hiding protected servers in the blockchain network and creating a creative routing strategy. DoS/DDoS attack detection AI models may be trained and executed via a trusted infrastructure without incurring a large cost by implanting the AI block into an intelligent contract utilising YODA. This revolutionary, affordable contract execution mechanism requires significant processing capability. According to a gaming study, DoS and DDoS attacks could be rather expensive.

How integrating ML with Blockchain enhances IoT security

One of the technologies that has seen the most adoption in recent years across a variety of applications is the IoT. In a variety of real-time scenarios, smart devices are connected via wired or wireless connections for communication, processing, computing, and monitoring reasons. The items lack computing power, have low memory, and are diverse. Due to the incompatibility of IoT devices with traditional-based security protocols, implementing the IoT system poses security and privacy issues.

Sensors used in the IoT create an enormous quantity of data, which is frequently refined at centralised servers. Centralised AI algorithms use IoT sensor data for a variety of purposes, including demand optimisation and production planning. IoT systems' main data collection points are edge devices like IoT network gateways. The training data has privacy issues in centralised AI training. Attackers can hack the entire data produced by the IoT system while they are practising. Conscious IoT data may reveal details about the real-time operations carried out in industries. This section will discuss how blockchain technology has advanced and how various ML algorithms are used to maintain IoT security.

Integration of Federated Learning with Blockchain for IoT Security

Federated Learning: Federated learning (FL), a method for processing dispersed data without compromising privacy, has recently become prominent. Fang et al. propose an architecture for integrating FL, IoT, and blockchain [17]. Algorithms are needed to process spatiotemporal data streams and unstructured data in complex industrial systems to provide the best decision support. For such systems, centralised single-database solutions are typically not recommended. Therefore, for secure big data analytical services, we consider distributed or federated systems that don't link blockchain with IoT.

FL is very suitable for use in cases in digital medicine when users are restricted from sharing data.

- Consider three scenarios incorporating AI/ML and potential AI assaults in digital medicine or healthcare supply chains. Then, a defence mechanism is provided using blockchain technology and fuzzy hashing.

In the first instance, healthcare professionals can use their faces to authenticate themselves to Healthcare Information Systems (HIS) [18]. In the second scenario, speech recognition might be used to authenticate patients, physicians, and pharmacists who connect to a dial-in telemedicine service. By entering their symptoms into an AI application, patients can use text recognition in the third example to find the best nearby healthcare provider.

Fig. 1.4.1 shows the structure of a private and secure FL for medical management. Based on federation architecture, a comprehensive model is generated on a shared server or solely end-to-end. The drug supplier, makers, and medical care sites will train the local AI network in this architecture, and these sites will securely transmit the taught models to an AI utility. This service is responsible for fusing blockchain verification with regional models to create a global model. Scalability, the elimination of a single point of failure, and the assurance of local model consistency are all advantages of the presented approach. This method maintains data flow and access control because it doesn't need a fully decentralised infrastructure.

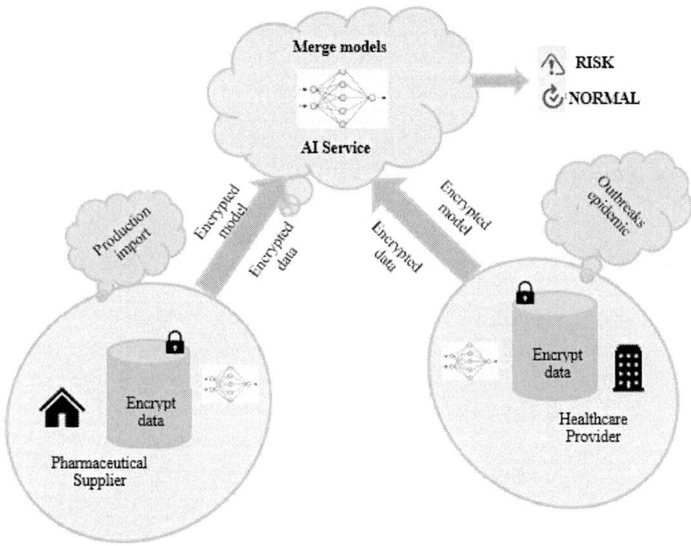

Fig.1.4.1. Role of FL in the Pharmaceutical supply chain

IoT security architecture

The study [19] designed a blockchain-enabled intelligent IoT architecture with AI to offer a workable technique for combining blockchain and AI for IoT with cutting-edge, contemporary methods and applications. This research [19] discusses the problems of privacy, accuracy, latency, and centralization by merging blockchain with AI in IoT applications. To address current issues, this design converges distributed edge blockchain networks at the fog layer, distributed cloud blockchain networks at the cloud layer, and peer-to-peer blockchain networks at the device layer.

The proposed architecture (Fig. 1.4.2) consists of four types of intelligence: device intelligence, edge intelligence, cloud intelligence, and fog intelligence. It makes use of this knowledge to demonstrate how blockchain technology and AI may be used together to address issues with massive information review, reliability, and systematisation in IoT services such as health maintenance, connected cities, and smart transportation.

Device Intelligence: The vast amount of data generated by device intelligence, which is composed of various IoT devices with blockchain and AI applications, is received by edge intelligence. **Edge Intelligence**: Base stations with AI capabilities that are connected to the blockchain at the network's edge make up edge intelligence. A variety of sensors are connected to each base station with edge intelligence and AI, and those sensors' traffic data is processed and analysed by the base station.

Fog Intelligence: The Fog Intelligence receives the processing data from the edge intelligence and mixes it with blockchain and many AI-enabled fog nodes.

Cloud Intelligence: Each blockchain-based AI-enabled fog node is connected to the edge intelligence's AI-enabled base stations and is in charge of processing data for cloud intelligence. Last, cloud intelligence has blockchain-connected, AI-based information centres that provide safe, massively distributed data for IoT services like smart healthcare, smart transportation, and so forth.

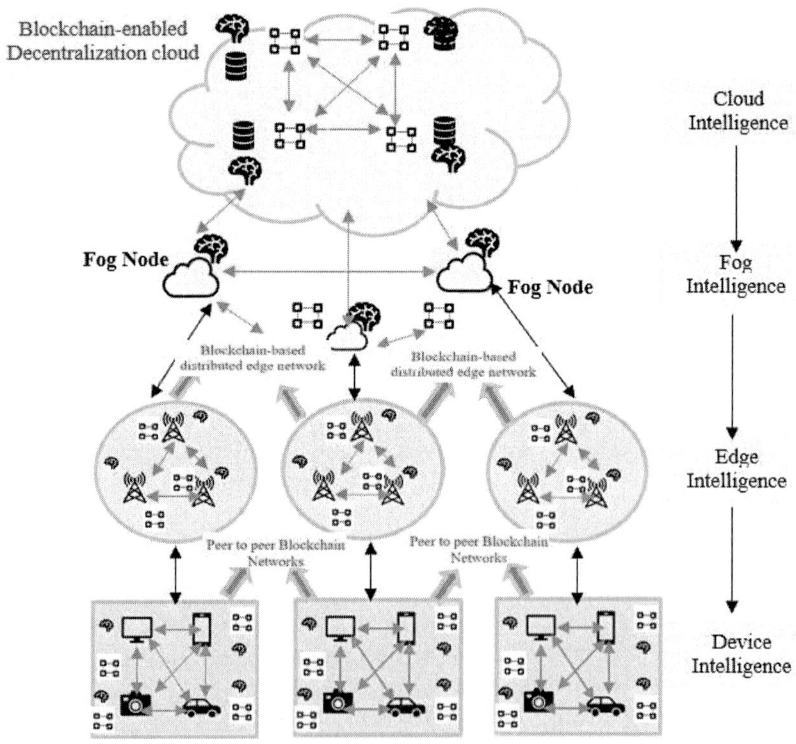

Fig.1.4.2. Secure IoT Architecture

Security analysis in IoT architecture

The six layers that make up an IoT platform are the physical, communication, link control, service, management, and application layers. Fig. 1.4.3 depicts the suggested architecture's orderly flow. In the suggested design, the physical layer correlates with devising intelligence, edge intelligence in the communication and link control layer, fog intelligence in the service and management layer, and cloud intelligence in the application layer. In cloud intelligence, the physical layer recognises information like temperature, location, pollution, weather, motion, and agriculture. Several sensor technologies, including radio frequency identification (RFID), barcodes, and infrared, were used to collect this data. This layer is susceptible to a variety of security issues and risks, including information transportation from one place to another, which makes it open to hackers. Blockchain and AI ideas are applied when a transaction is made using bitcoin, litecoin, or Ethereum to mitigate these issues. The communication layer, which acts as a channel for data to be delivered from one unit to another, receives the obtained data. Several cutting-edge technologies are used, including Wi-Fi, Zigbee, radio, and infrared waves. Point-to-point networks leverage blockchain and AI technologies, and pervasive wideband is utilised for encryption and verification. Privacy and security issues exist with this layer.

The IoT blockchain and AI convergence need consensus mechanisms for scalability and security. It provides distribution and decentralisation tools. This layer is utilised for decentralised information storage. After receiving the data from storage, the service layer provides crucial services for IoT applications, such as virtual entity resolution, database support, service configuration and operation, decision support, and IoT service observation. For secure verification and confirmation in this layer, the intersection of blockchain and AI for IoT uses distributed cloud and intelligent storage, micro-servers, and smart contracts. The application layer may manage data, software, standards, and infrastructure across networks thanks to this data flow to the management layer. It provides digital specification, hash function, micro-server, and design code for enciphering, integrating blockchain and AI for IoT. Not to mention that information shared with the application layer enables global administration of IoT applications used in smart cities, smart cars, smart healthcare, smart farming, intelligent transportation, and other applications.

Utilising other technologies like ML, programmable learning, deep learning, analytics intelligence, and ML, blockchain is integrated with AI for IoT. Some of the critical characteristics are depicted in Fig. 1.4.4.

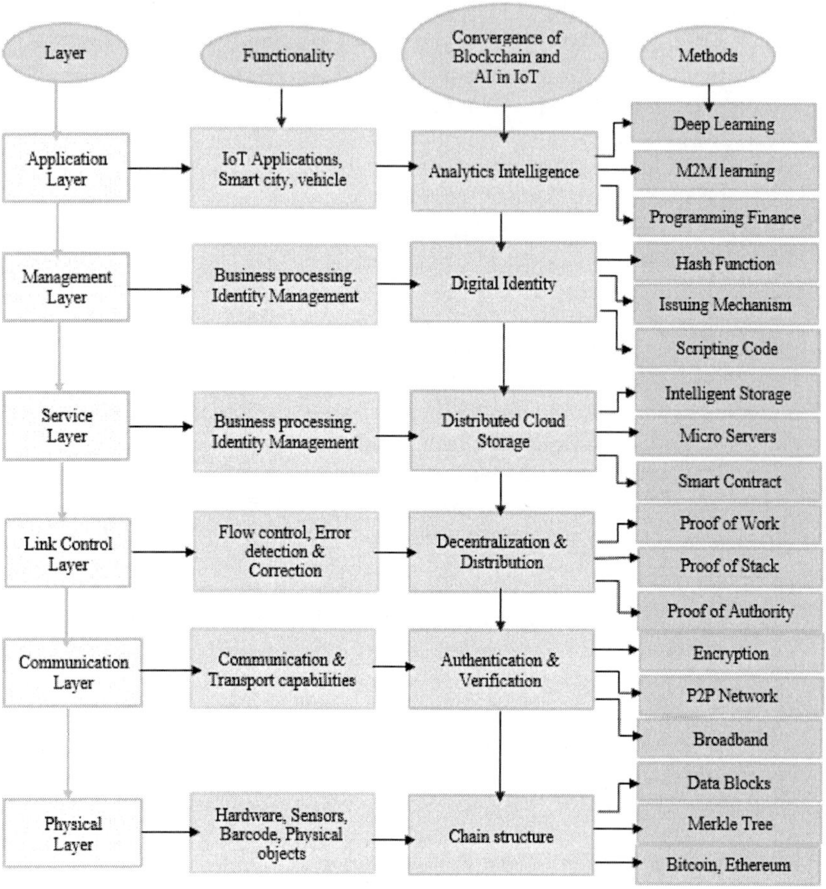

Fig.1.4.3. Characteristics of Block Intelligence

Blockchain-supported information protection for AI applications in 6G networks

As terrestrial wireless and satellite communications merge, it is projected that B5G networks will provide a completely connected environment. The B5G architecture's major goal is to use AI to support the cautious and quick development of system utilities. Intelligent services process a lot of data naturally through storage, computing, and evaluation, which leaves the data open to alteration or contamination by outsiders. Blockchain-based information security in B5G networks is discussed in this section.

Data security in cyberspace

Singh et al. propose the computer network architecture by integrating the following essential elements [20]:

- Sharing data on the blockchain with ownership guarantees;
- The platform for secure computing powered by big data and AI creates intelligent, dynamic security rules; and
- A system for exchanging value to pay for security services.

The proposed secure network's overall architecture is shown in Figure 1.5.1. Blockchain-based networking links the network's nodes together. Nodes use blockchain-based techniques in the network to communicate with one another and reach agreements. They collaborate while executing classy contracts in the interval. To obtain consensus on the node level or the outcomes of intelligent-contract implementation, each node has a blockchain registry that it uses for matching the levels with other nodes. Access control and data storage modules are included in secure-network nodes for secure data storage. The Operation Support System (OSS) module on secure network nodes enables AI-based secure computing (ASC) to develop knowledge and security rules from data.

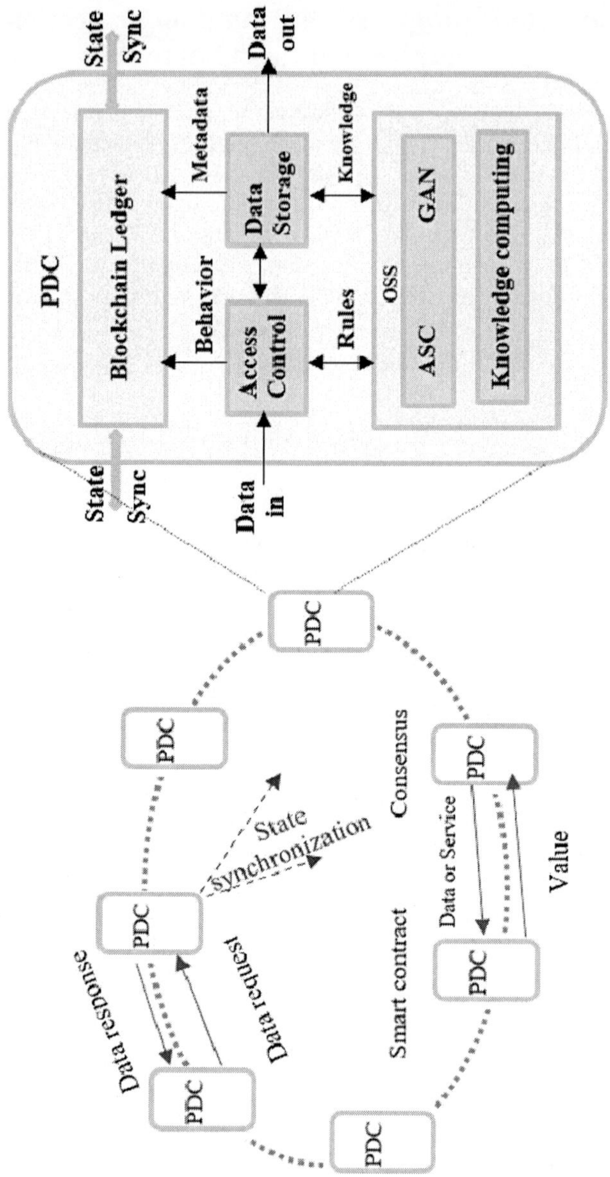

Fig.1.5.1. Secure Network Architecture

Data sharing: A secure network uses a blockchain-based data-sharing protection mechanism and the Private Data Centre (PDC) for data security. PDC uses state-of-the-art architectural and technical methods to provide physical protection for data while operating AI-based OSS. One crucial feature of PDC is universal data access control.

Secure Computing: ASC can incorporate the Generative Adversarial Network (GAN) module at the beginning of secure computing to provide a safe and intelligent OSS for PDC and produce more potent and developing security rules. The GAN module of ASC can ascertain the current security rules of a PDC and then produce malicious access requests for certain secret data that "look valid" to confuse the OSS of a PDC and prevent it from classifying the request as either unlawful or not. Fake data access requests choose to compete with the GAN module's highly developed and secure OSS after a protracted cycle of generating and categorising requests.

AI-Blockchain-based data security

In the research work [21], data security is enhanced with FL. Model State, together with the idea of channels and data transit. It puts "Device Register," "Models Update," "Models Consensus," and other ideas into practise using the FL-based MEC-enabled blockchain network. For ML on global models, the FL-based blockchain principle generated a unique channel for each model. Instead of saving the models in a single data store, the references will be kept on a highly secure distributed server, a global model tree maintained by Merkle Patricia.

This platform is not ideal for large-scale data transfers since the user devices depend on the hardware at the network's edge. They want to improve data processing, latency, and storage platform requirements in the future. The FL-based blockchain, however, has a framework for a set of data mining nodes and equipment.

A distributed ML approach called FL allows testing ML models on personal, decentralised data. Because ML models are trained locally, FL guarantees data privacy while eliminating the transfer of huge amounts of training data via networks. FL is utilised to predict words in programmes like Google Keyboard, hospital healthcare to predict hospital stays and mortality, and the telecommunications industry at network edges and cores. A communication between an FL server and two FL clients is formed in Umer and Seon's paper to illustrate a scenario [22]. Blockchain technology underpins the data exchange system for FL in Fig. 1.5.2, and every peer node in the blockchain network has a compatible reproduction of the blockchain. For an FL task, the server and clients can publish the illustration of their ML models on the blockchain using the metadataPublish() method. The FL server and clients can explore the metadata stored in their blockchain copy using metadataGet(). If the pertinent ML model's information can be found, dataGet will then send it straight from the server to the clients. DataGet() establishes a secure network connection and enables information transmission between servers and clients.

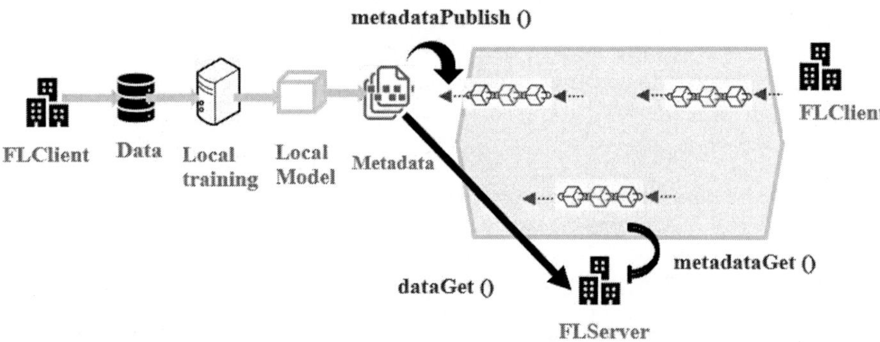

Fig.1.5.2. Data exchange using Blockchain

Blockchain and AI-empowered secure6G Architecture

As discussed in previous sections, blockchain can create a decentralised environment where resources are shared safely. AI can tackle issues with complex, ambiguous, and time-variant properties.

Information-Centric Network

The article [23] proposed an intelligent and safe architecture for future-generation wireless communication incorporating AI and blockchain, as shown in Fig. 1.6.1, to allow dynamic and trustful resource allocation. Spectrum splitting, content caching, energy exchanging, and compute offloading are just a few examples of the resources that the suggested architecture may manage and orchestrate securely and intelligently. The authors then create a caching strategy for boosting the system utilising the cutting-edge DRL algorithm, using blockchain to provide a trustworthy environment for content caching.

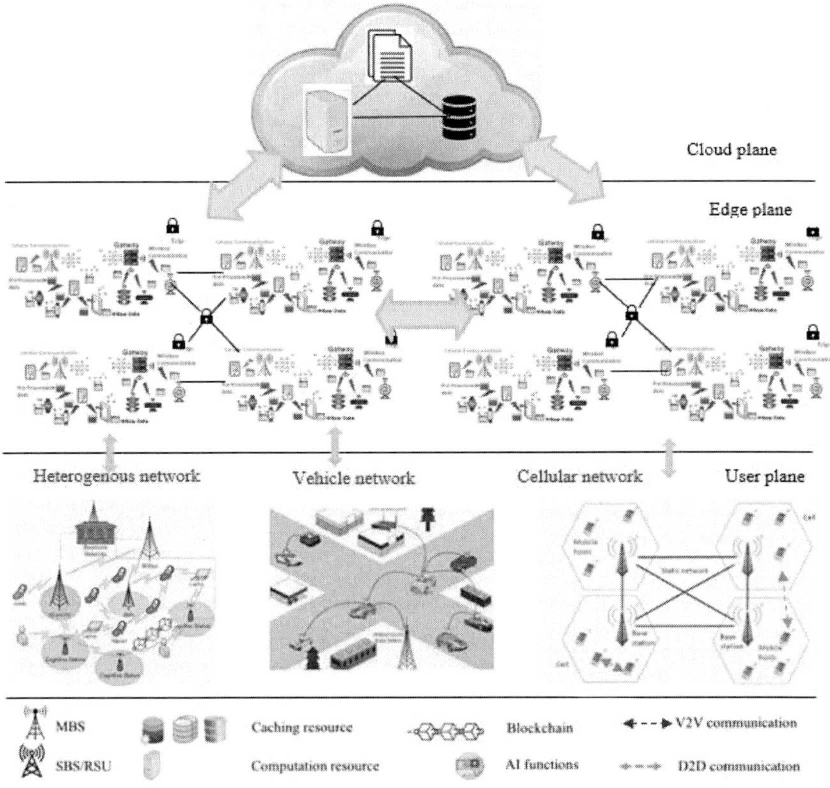

Fig.1.6.1. Secure Intelligent Future Network

The three layers of architecture are the cloud, edge, and user layers. The cloud plane has many machines with intense processing, caching, and calculation capabilities. This layer can leverage a global perspective and cutting-edge methods like information extraction and huge amounts of information (big data) to shift network states. Coordination from instinctive network functioning to driven network management happens by anticipating a few incidents or reallocating a few resources.

Because of their powerful data processing capabilities and plentiful cache resources, cloud servers can run delay-tolerant practises and reserve data that is huge in dimension or unpopular. There is also a main control in the cloud layer. The key control, which is furnished with secured technology, controls mobile devices, smart automobiles, large base stations, microbase stations (MBSs), and roadside units (RSUs).

At the network edge, MEC servers and blockchains are put in close-by network infrastructures (like MBSs and RSUs). Network infrastructures can provide radio interfaces for mobile devices and vehicles to guarantee flawless coverage and quick wireless communication. Thanks to MEC servers' computation, caching, and AI capabilities, essential material, like the most recent information and warnings, can be cached at the network edges. This makes it possible for them to run time- and calculation-sensitive applications. Blockchain can maintain a distributed registry and trace every wireless system transaction to improve the security and privacy of the wireless ecosystem. The scenario supports vehicle-to-RSU and vehicle-to-vehicle (V2V) communication in vehicular networks. Vehicles can communicate with one another to transfer information or energy. Because they may not be able to trust one another, cars must employ pseudonyms when transmitting content or energy to increase security and privacy protection. To do this, a distributed ledger is maintained and implemented on the blockchain on RSUs to expand the safety and seclusion of the wireless ecosystem.

The work [24] provides a more thorough overview of the next generation of wireless networks (Fig. 1.6.2) and discusses how new scientific approaches, which are popular investigation areas, may be applied to boost the network's security. For example, using blockchain concepts to deliver security-critical services safely, rapid security processes at numerous network positions, fog nodes, safety automation and resolution, which could need AI techniques, and so forth are a few examples.

Numerous security approaches could be used in these situations to ensure security. For example, portable software-based security features, service- or application-based security, effective network slicing-based isolation, and AI-based dynamic security evaluations in multiple network circumferences. The following subdivisions examine how upcoming wireless communication standards can be safeguarded using cutting-edge technologies like security softwarisation and virtualisation, AI, and blockchain technologies.

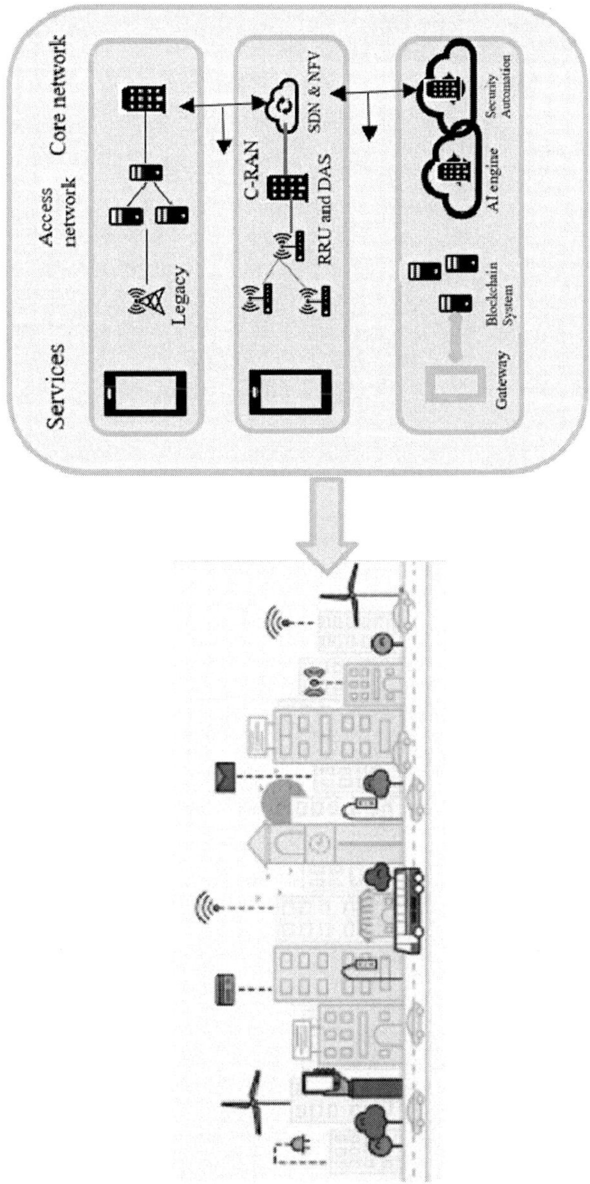

Fig.1.6.2 Secure architecture for future cities

Fig. 1.6.3 depicts a potential 6G design [12] incorporating the Blockchain-Radio Access Network (B-RAN). The Edger Data Centre (ESDC) comprises robust baseband units (BBUs) and edge servers using cutting-edge technology like blockchain, AI, and big data. As an improved 5G evolved Node B (eNodeB), an ESDC works with numerous Remote Radio Units (RRUs) to provide wireless applications, a variety of local applications, the network market, and other services. Blockchain confirms internal safety within an ESDC and entitles it to several vital activities like highly safe access control, tracing, and guidance of mobile devices using AI, big data, and encryption. Blockchain facilitates secure and reliable interactions between the numerous EDSCs connected by optical cables for fast data transmission. Along with ESDCs, blockchain also links substantial numbers of different end devices, gateways, and core systems. It manages and defines the licences and responsibilities of every participant and, in the end, realises system architecture via on- and offline sensible agreements.

Especially, B-RAN can establish confidence among IoT elements and access points in harsh conditions by using the fundamental blockchain in an IoT scenario. It can also provide a future strategy for IoT and IoE in a multi-operator network. Building trust prevents potentially selfish behaviours among unreliable devices and promotes cooperation between IoT networks.

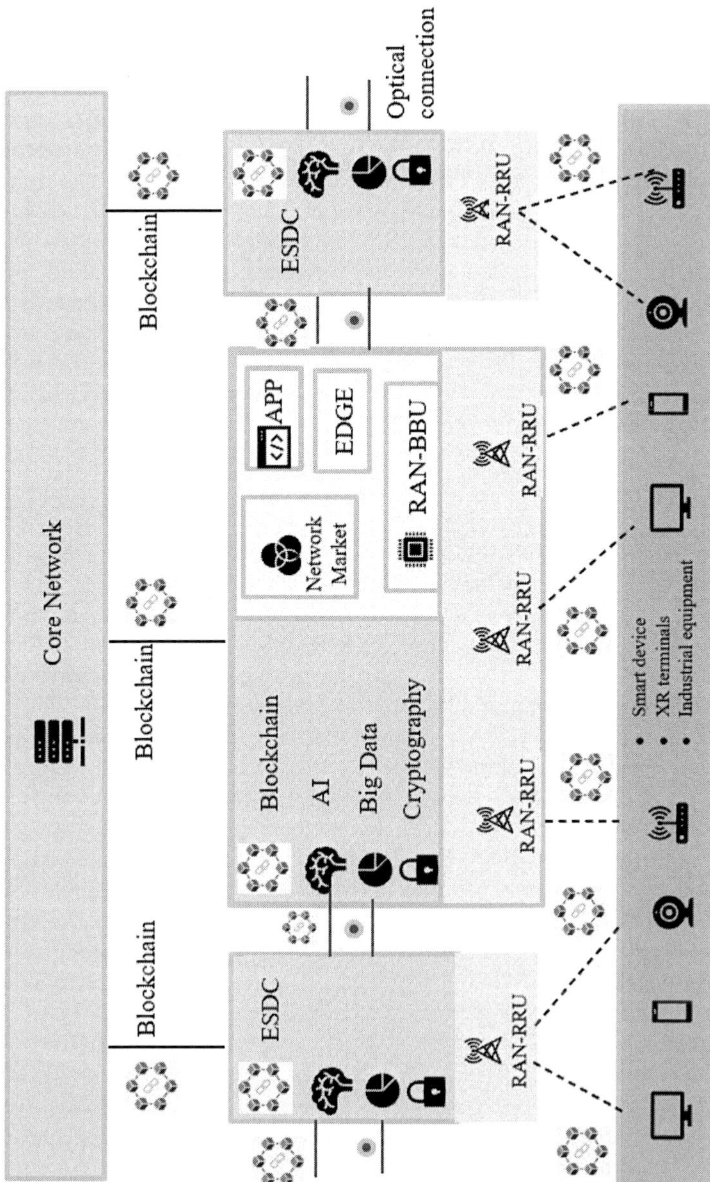

Fig.1.6.3.6 G-BRAN Architecture

By reconfiguring numerous independent system devices into a cooperative diverse-agent system dependent on the blockchain, B-RAN may systematically combine and use cross-layer resources, such as spectrum, access points, IoT units, and user information. Thanks to effective incentive mechanisms in B-RAN, IoT devices can access resources and services across networks; therefore, they are not limited to servicing only one subscriber. Another integrated B-RAN product that uses blockchain technology, MEC, offers multi-party resource scheduling for a widespread network while maintaining user privacy and information protection. Without the need for middlemen, B-RAN provides flexible, straight connectivity between users and MEC servers of various operators in the network. By utilising MEC members' storage and computational resources, B-RAN may efficiently design resource sharing and scheduling while lowering vacancy and redundancy in the network administration.

Conclusion

Wireless communication networks have advanced to connect every aspect of life in 5G virtually. Throughout this progression, the security environment has also altered, evolving from simple eavesdropping to a variety of strikes against mobile units, system architecture, and applications. B5G can leverage innovative trends like massive multiple input multiple output [25] (massive MIMO), SDNs, NFV, and complex cloud computing ideas to incorporate the latest IoT devices and assistance into the system. The network security landscape may grow more complex due to the inherent security problems these technologies provide. As a result, we have discussed in this chapter the security concerns that may exist in B5G networks. The development of numerous units, applications, and new association mechanisms necessitates the investigation of a novel security solution for efficient and secure communication. As a result, we have covered the security concerns that are raised by the various B5G network components and technologies in depth and have offered several potential security guidelines, techniques, and fixes. Key issues related to trust in wireless networks that hinder the evolution of B5G networks into safer and more efficient 6G networks have been discussed. Innovative communication services and technology usage will also raise security issues and challenges. But tackling these problems from the start of design to deployment will lessen the possibility of protection and confidentiality slippage.

References

2. Porambage, Pawani, Gürkan Gür, and Diana Pamela Moya Osorio, Madhusanka Livanage, and Mika 2021 "6G security challenges and potential solutions." In *2021, Joint European Conference on Networks and Communications and 6G Summit (EuCNC/6G Summit)*, 622–627.
3. Vasuki, and Vijayakumar Ponnusamy. "Latest Wireless Technologies towards 6G." In *2021 International Conference on Computer Communication and Informatics (ICCCI)*, pp. 1–4, IEEE, 2021.
4. Ahmad, S. Shahabuddin, T. Kumar, J. Okwuibe, A. Gurtov, and M. Ylianttila 2019. "Security for 5G and Beyond," in *IEEE Communications Surveys & Tutorials*, vol. 21, No. 4: 3682–3722. DOI: 10.1109/COMST.2019.2916180.
5. Li, Yannan, Yong Yu, Willy Susilo, Zhiyong Hong, and Mohsen Guizani, 2021. "Security and privacy for edge intelligence in 5G and beyond networks: Challenges and solutions." *IEEE Wireless Communications*, 28: 63–69.
6. Malathi, D., Vijayakumar Ponnusamy, S. Saravanan, D. Deepa, and Tariq Ahamed Ahanger "A Design Framework for Smart Ration Shops Using Blockchain and IoT Technologies" *Intelligent Automation & Soft Computing,* 32, no. 1 (2022).
7. Han, Sicheng, and Xiaorong 2019. "Blockchain-based spectrum sharing algorithm." In the *2019 IEEE 19th International Conference on Communication Technology (ICCT)*, 936–940
8. Kotobi, Khashayar, and Sven G. Bilen. 2018. "Secure blockchains for dynamic spectrum access: A decentralised database in moving cognitive radio networks enhances security and user experience." *IEEE vehicular technology magazine*, No. 1: 32–39.
9. Zhou, Zhenyu, Xinyi Chen, Yan Zhang, and Shahid 2020 "Blockchain-empowered secure spectrum sharing for 5G heterogeneous networks" *IEEE Network*, No. 1: 24-31.
10. Fan, Xin, and Yan Huo. 2020. "Blockchain-based dynamic spectrum access of non-real-time data in cyber-physical-social systems." *IEEE Access No. 8*: 64486–64498.
11. Maksymyuk, Taras, Juraj Gazda, Longzhe Han, and Minho 2019 "Blockchain-based intelligent network management for 5G and beyond" In *2019, 3rd International Conference on Advanced Information and Communications Technologies (AICT)*, 36–39

12. Yang, Dana, Seohee Yoo, Inshil Doh, and Kijoon Chae 2021. "Selective blockchain system for secure and efficient D2D," *Journal of Network and Computer Applications No. 173*: 102817.
13. Wang, Jiaheng, Xintong Ling, Yuwei Le, Yongming Huang, and Xiaohu You 2021. "Blockchain-enabled wireless communications: A new paradigm towards" *National Science Review*, No. 9: nwab069.
14. Tyagi, Amit Kumar, S. U. Aswathy, and Ajith Abraham 2020. "Integrating blockchain technology and artificial intelligence: synergies, perspectives, challenges, and research directions." *Journal of Information Assurance and Security*, 5: 1554
15. Li, Weiwei, Zhou Su, Ruidong Li, Kuan Zhang, and Yuntao 2020 "Blockchain-based data security for artificial intelligence applications in 6G networks" *IEEE Network*, No. 6: 31–37.
16. Pan, Qianqian, Jun Wu, Jianhua Li, Wu Yang, and Zhitao Guan "Blockchain and AI empower a trust- and information-centric network for beyond 5G." *IEEE Network*, No. 6: 38–45
17. Fang, Liming, Bo Zhao, Yang Li, Zhe Liu, Chunpeng Ge, and Weizhi Meng. 2020. "Countermeasure based on smart contracts and AI against DoS/DDoS attack in 5G." *IEEE Network*, No. 6: 54–61.
18. Unal, Devrim, Mohammad Hammoudeh, Muhammad Asif Khan, Abdelrahman Abuarqoub, Gregory Epiphaniou, and Ridha Hamila 2021: "Integration of federated machine learning and blockchain for the provision of secure big data analytics for the Internet of Things." *Computers & Security*, 109: 102393
19. Ponnusamy, Vijayakumar, Vasuki, J. Christopher Clement, and P. Eswaran, "AI-Driven Information and Communication Technologies, Services, and Applications for Next-Generation Healthcare Systems." *Smart Systems for Industrial Applications* (2022): 1-32.
20. Singh, Sushil Kumar, Shailendra Rathore, and Jong Hyuk Park. 2020. "Blockiotintelligence: A blockchain-enabled intelligent IoT architecture with artificial intelligence." *Future Generation Computer Systems*, No. 110: 721–743
21. Wang, Kai, Jiaqing Dong, Ying Wang, and Hao 2019. "Securing data with blockchain and AI nd Hao 2019. "Securing data with blockchain and AI" *Ieee Access No.* 777981-77989.
22. Majeed, Umer, and Choong Seon 2019. "FL chain: Federated learning via MEC-enabled blockchain network." In 2019, the *20th Asia-Pacific Network Operations and Management Symposium (APNOMS)*, 1-4.

23. Zhang, Qiong, Paparao Palacharla, Motoyoshi Sekiya, Junichi Suga, and Toru 2020 "A blockchain-based protocol for federated learning" In *2020, IEEE will hold the 28th International Conference on Network Protocols (ICNP)*, 1-2.
24. Dai, Yueyue, Du Xu, Sabita Maharjan, Zhuang Chen, Qian He, and Yan 2019. "Blockchain and deep reinforcement learning empowered intelligent 5G and beyond." *IEEE Network*, No. 3: 10–17.
25. Ahmad, Ijaz, Shahriar Shahabuddin, Tanesh Kumar, Jude Okwuibe, Andrei Gurtov, and Mika 2019. "Security for 5G and beyond." *IEEE Communications Surveys & Tutorials*, No. 4: 3682–3722
26. Ponnusamy, Vijayakumar, and Malarvihi, "Hardware impairment detection and prewhitening on MIMO precoder for spectrum sharing." *Wireless Personal Communications* 96 (2017): 1557–1576